NATO ASI Series

Advanced Science Institutes Series

A series presenting the results of activities sponsored by the NATO Science Committee, which aims at the dissemination of advanced scientific and technological knowledge, with a view to strengthening links between scientific communities.

The Series is published by an international board of publishers in conjunction with the NATO Scientific Affairs Division

A Life Sciences B Physics	Plenum Publishing Corporation London and New York
C Mathematical and Physical Sciences D Behavioural and Social Sciences E Applied Sciences	Kluwer Academic Publishers Dordrecht, Boston and London
F Computer and Systems Sciences G Ecological Sciences H Cell Biology I Global Environmental Change	Springer-Verlag Berlin Heidelberg New York London Paris Tokyo Hong Kong Barcelona

NATO-PCO DATABASE

The electronic index to the NATO ASI Series provides full bibliographical references (with keywords and/or abstracts) to more than 30000 contributions from international scientists published in all sections of the NATO ASI Series. Access to the NATO-PCO DATABASE is possible in two ways:

- via online FILE 128 (NATO-PCO DATABASE) hosted by ESRIN, Via Galileo Galilei, I-00044 Frascati, Italy.
- via CD-ROM "NATO-PCO DATABASE" with user-friendly retrieval software in English, French and German (© WTV GmbH and DATAWARE Technologies Inc. 1989).

The CD-ROM can be ordered through any member of the Board of Publishers or through NATO-PCO, Overijse, Belgium.

Series F: Computer and Systems Sciences Vol. 78

The ASI Series Books Published as a Result of Activities of the Special Programme on ADVANCED EDUCATIONAL TECHNOLOGY

This book contains the proceedings of a NATO Advanced Research Workshop held within the activities of the NATO Special Programme on Advanced Educational Technology, running from 1988 to 1993 under the auspices of the NATO Science Committee.

The books published so far as a result of the activities of the Special Programme are:

Vol. F67: Designing Hypermedia for Learning. Edited by D. H. Jonassen and H. Mandl. 1990.

Vol. F78: Integrating Advanced Technology into Technology Education. Edited by M. Hacker, A. Gordon, and M. de Vries. 1991.

Series F: Computer and Systems Sciences, Vol. 78

Integrating Advanced Technology into Technology Education

Edited by

Michael Hacker

New York State University
Department of Education
One Commerce Plaza
Room 1619
Albany, NY 12234, USA

Anthony Gordon

Education Inspectorate
The Kingston Centre
Harrowby Street
Stafford, ST16 3TU, UK

Marc de Vries

Pedagogical Technological College
P.O. Box 826
5600 AV Eindhoven, The Netherlands

Springer-Verlag
Berlin Heidelberg New York London Paris Tokyo
Hong Kong Barcelona Budapest
Published in cooperation with NATO Scientific Affairs Division

Proceedings of the NATO Advanced Research Workshop on Integrating Advanced Technology into Technology Education, held in Eindhoven, The Netherlands, October 9–12, 1990.

ISBN-13: 978-3-642-76770-8 e-ISBN-13: 978-3-642-76768-5
DOI: 10.1007/ 978-3-642-76768-5

Library of Congress Cataloging-in-Publication Data
NATO Advanced Research Workshop on Integrating Advanced Technology into Technology Education (1990: Eindhoven, Netherlands) Integrating advanced technology into technology education/edited by Michael Hacker, Anthony Gordon, Marc de Vries. p. cm.—(NATO ASI series. Series F, Computer and systems sciences; vol. 78) "Proceedings of the NATO Advanced Research Workshop on Integrating Advanced Technology into Technology Education, held in Eindhoven, The Netherlands, October 9–12, 1990"—T.p. verso. "Published in cooperation with NATO Scientific Affairs Division." Includes index.

1. Technical education—Congresses. 2. High technology—Congresses. I. Hacker, Michael. II. Gordon, Anthony. III. Vries, Marc de. IV. North Atlantic Treaty Organization. Scientific Affairs Division. V. Title. VI. Series: NATO ASI series. Series F, Computer and system sciences; vol. 78. T62.N33 1990 607.1—dc20 91-22315

Softcover reprint of the hardcover 1st edition 1991

Camera-ready copy provided by the editors

45/3140-543210 – Printed on acid-free paper

Foreword

Technology is a dynamic field in which new developments take place continuously. Education traditionally lacks behind the latest developments in this subject area. Therefore it is necessary to consider updating education regularly. The NATO Advanced Research Workshop "Integrating Advanced Technology into Technology Education" discussed necessary changes in technology education caused by new developments in technology. It was held in Eindhoven, The Netherlands, October 9-12, 1990. Our impression is that there was a fruitful interaction between educationalists, policy makers, and industrialists. These groups all have their own way of approaching technology education and the conference gave them an opportunity for sharing ideas that come from these various approaches. The participants felt that the issue of integrating advanced technology into technology education was quite relevant. There was a high degree of commitment in presentations and discussions. This led to the formulation of a number of recommendations to people in NATO countries who are involved in technology education. We as organizers appreciate the contributions of many people to this conference: NATO for financing it; Dr. Thomas Liao in particular for stimulating both NATO and us to realize it; the presenters for setting the scene for discussions; all participants for sharing ideas and perspectives; and finally Springer-Verlag for publishing the proceedings.

April 1991

Michael Hacker
Anthony Gordon
Marc de Vries

Introduction

From the main theme of the conference -- Integrating Advanced Technology into Technology Education -- we distilled six questions:

1. What are "advanced" technologies and what makes them relevant to technology education?
2. What is the state of affairs in contemporary technology education?
3. How does the integration of advanced technology affect relationships between technology education and other curriculum areas?
4. What strategies are there for integrating advanced technology into technology education?
5. What are criteria for selecting advanced technologies to be integrated into technology education?
6. How can international collaboration stimulate the integration of advanced technology into technology education?

These questions were addressed in the six parts of the conference program and they have been dealt with in the same order in these proceedings. Each part of the proceedings consists of some introductory remarks and a brief summary of the presentations, followed by the full text of the presentations.

TABLE OF CONTENTS

IV. Implementation Strategies

V. Selecting Advanced Technologies to be Integrated

VI. Fostering International Collaboration

I. The Nature and Relevance of Advanced Technology

When discussing the integration of advanced technology into technology education, it is necessary to define what one is talking about. Therefore the first issue to be addressed was: What do we mean by "advanced" technology? In other words, what makes a technology "advanced"? The whole concept of technology is very complex and it can hardly be expected that this conference would yield a complete, one-sentence definition of "advanced technology." Therefore a more realistic aim was set: to search for general features that determine the "advancedness" of technology. Together with the issue of the nature of advanced technology goes its importance: why at all do we want to discuss the possible role of advanced technology in technology education?

In his opening address, **Dr. Feye Meijer** (Philips Corporation, the Netherlands) emphasized the need for confronting our children with new developments in technology. In their daily lives they are constantly using advanced technological products, and to do that well and wisely they should have some understanding of the way in which these products are developed.

Dr. Dennis Swyt (National Institute of Standards and Technology, USA) in his keynote paper made evident that important changes in the labor market are taking place that will require a technical-professional work force in the coming decades. In his definition of advanced technology he related to the new and powerful capabilities that go with the advanced technologies and the fact that they are based on newly recognized lawful behavior in nature.

Mr. Jean-Jacques Duby (IBM Corporation, France) gave a survey of the various advanced technologies that can be found in industrial practice today. His examples came from mathematical, physical, and bio-related technology. He also made a plea for balancing advanced and traditional technologies in education, because these are both relevant in industry.

The outcome of the group discussions was that the nature of advanced technology was related to the role of science in technology. In advanced technology, scientific knowledge is used intensively. Science also has a methodological influence on advanced technology in the sense that it enhances analysis and abstraction in technological problem solving. The relevance of advanced technologies was not in the first place seen as the introduction of a number of trendy new specific and highly specialized "high-tech" issues, most of which are still in a laboratory phase. The relevance was rather seen in the stimulus for teaching and learning general basic concepts and principles. These generalizations will bring technology education beyond the level of teaching and learning specific knowledge and skills that are soon outdated. Secondly, introducing advanced technologies into technology education is necessary to give the students an understanding of the technologies they see around them. For a good conceptualization it is necessary to have a balance of traditional and advanced technologies in technology education.

In the third place it is relevant for technology education as an orientation on technological careers to be updated, so that students get a good impression of the contemporary situation. The industrial representatives had a special meeting during the conference in which they discussed this aspect, and in their report they underlined the need for good relationships between technology education and industry. Not only technological professions, but also all professions are influenced by the introduction of advanced technologies. Technology education should make students aware of this.

Investing in Education to Meet a National Need for a Technical-Professional Work Force in a Post-Industrial Economy

Dennis A. Swyt

National Institute of Standards and Technology, USA

Abstract: Work force data show that the U.S. is shifting from blue-collar production to technical-professional service as the basis of its post-industrial economy. One factor contributing to this shift is the continuous rise in the productivity of production, both agricultural and industrial, based on increased use of technology, *i.e.*, machines, measured in dollar-value of equipment per worker. The lesson of increased effectiveness with increased capital per person has not been lost on U.S. service industries, which have for the last decade led the U.S. economy in capital spending. In contrast, what may be the most important service industry -- basic public education -- has undergone a steep 15-year-long decline in its growth in capital investment per student. This suggests that means to focus on capital spending in basic education, such as instituting capital-equipment budgets, are in order.

Key words: work force, productivity, service economy

Introduction

During the 19th and early 20th centuries, the US basic educational system evolved to meet the needs of the U.S. economy, including those of corporate and governmental institutions, by providing to the future members of the work force the basic skills of literacy and numeracy represented by the proverbial three R's.

In this prior era, the majority of the U. S. work force made its living in labor-based production occupations in the agriculture and manufacturing sectors and agricultural and industrial arts programs provided the principal technology oriented, livelihood-relevant instruction.

As this paper will show, to a large degree today and increasingly more so in the future, the U.S. work force makes its living in a fundamentally different way -- rapidly shifting out of goods-production industries and physical-labor occupations into service-providing industries and technical-professional occupations. Concurrently, use of technology -- represented by an increasing capital intensity of service as well as production industries -- is dramatically

NATO ASI Series, Vol. F 78
Integrating Advanced Technology into Technology Education
Edited by M. Hacker et al.

increasing. Taken together, these sectoral, occupational, and capital-intensity trends suggest that a rethinking of the role and level of investment in technology in education is in order

The Shift from Production to Service Sectors

Figure 1 shows for the period from 1820 to 1995 the fractions of the U.S. work force within the manufacturing-and-mining, agricultural, and service sectors [1]. Illustrated are: (1) the smooth evolution out of the era in which agriculture formed the dominant component of the overall work force, decreasing from as high as 90% in colonial times to 2.5% by 1995; (2) the transitions both into and out of the era in which the manufacturing work force dominated, beginning with the sustained increase during the period from 1830 to 1900, through the plateau-like level of 25% for the period between 1900 and 1975, and after 1975, the on-set of a decline; and (3) the continuous increase in the size of the service sector, now representing about 70% of the work force.

The recent onset of decline in the relative numbers of workers in the manufacturing-mining sector presents a clear demarcation point for the move of the U.S. economy and society into a "post-industrial" era; note, however, that the move into a "service economy" began, in a real sense, with the beginning of industrialization itself.

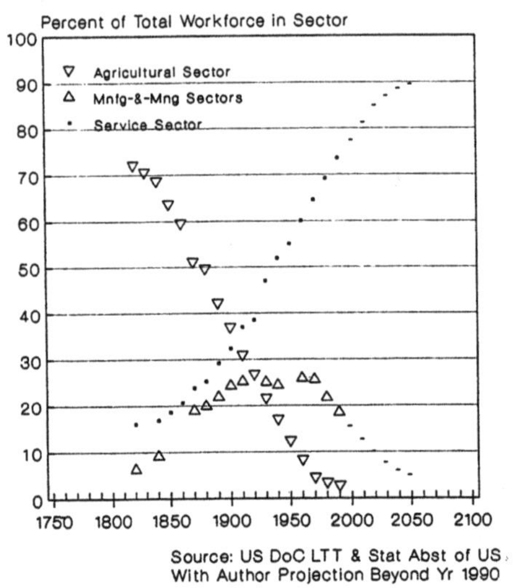

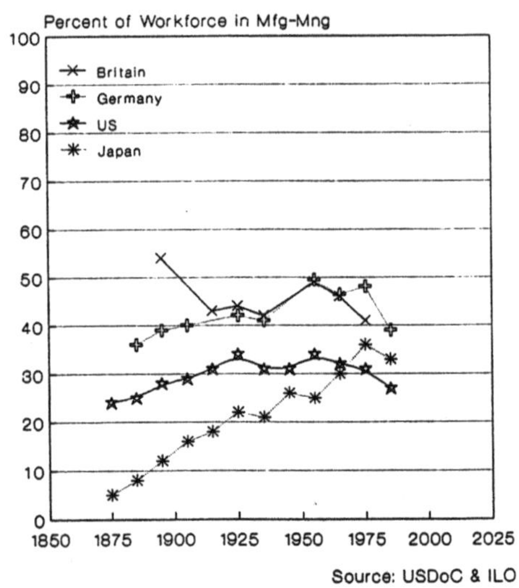

Figure 1. Movement of US work force from agriculture into and out of manufacturing concurrent with rise of service sector

Figure 2. Movement of industrialized nations' work forces out of manufacturing-mining sector led by US

Note that as Figure 2 indicates, in terms of changes in the relative size of the manufacturing work force, while the U.S. followed Britain and Germany into the Industrial Revolution (followed, in turn, by Japan), the U.S. now leads all these industrialized nations into the post-industrial era [3].

Shift from Physical-Labor to Technical-Professional Occupations

Figure 3a shows for the period 1900 to the present the distribution of the total U.S. work force among four occupational categories [1]: *physical-production* occupations including laborer, operative, precision-production, and craft; *physical-service* occupations including, for example, hospital orderlies, parking-lot attendants, custodians, security guards, and fast-food workers; *managerial-administrative* occupations consisting managerial, administrative, clerical, and sales; and *technical-professional* occupations including, for example, engineers, scientists, financial analysts, nurses, doctors, accountants, technicians, and para-professionals of all types. Apparent in this time-based representation of data are, for example, the continuous decline in the fraction of the work force in the production occupations and the increase in the managerial-administrative.

In Figure 3b, the same occupational data are represented in a different manner. Along the vertical axis (PP-TP) is plotted the difference between the respective percentages of physical-

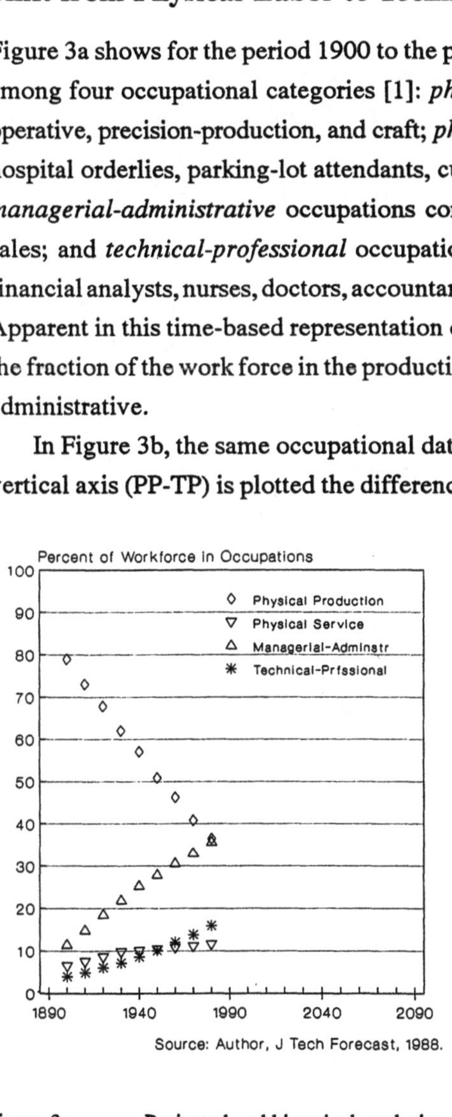

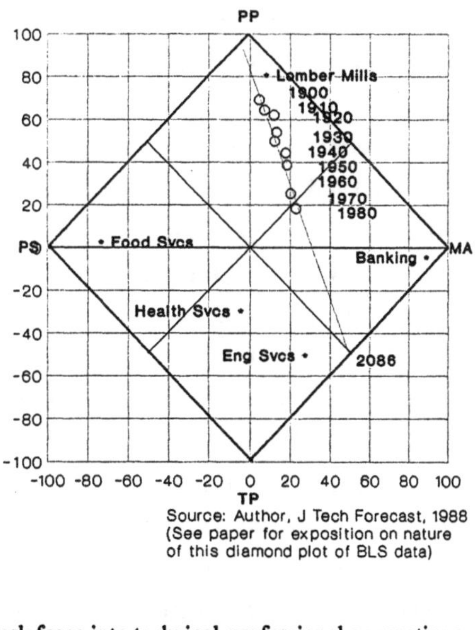

Figure 3a. **Projected and historical evolution of US work force into technical-professional occupations**
Figure 3b. **Historical and projected evolution of US work force along its 1900-1980 linear-fit time trend**

production (PP) and technical-professional (TP) workers; along the horizontal axis (PS-MA) is plotted, the difference between the respective percentages of managerial-administrative (MA) and physical-service (PS) workers. The sum of the four categories is 100%.

Superimposed on the plot for reference is a grid dividing the diamond diagram into four sectors representing, in effect, a "blue-collar" physical-production quadrant at the top, a "tan-collar" physical-service quadrant at the left, a "white-collar "managerial-administrative quadrant at the right, and a "no-uniform" technical-professional quadrant at bottom. Also included are single-year data for specific industrial sectors selected for their extreme characteristics: for example, (1) lumber mills -- at the top -- are almost purely physical-production; (2) banks -- at the right -- are almost purely managerial-administrative; (3) engineering services -- near the bottom -- are most highly technical-professional; while (4) food services -- at the left -- are almost purely physical-service.

Apparent in Figure 3b is the motion of the U.S. work force as a whole along a definite trajectory, one indicated by the least-square-fitted line and having two major features. The first is the historical crossing in 1975 from the upper-left domain, that of the physical-production and physical-service occupations, into the lower-right domain, that of the managerial-administrative and technical-professional. This "physical work to mental work "transition is made explicit in Figure 3c, in which the sums of the physical-work occupation (PP and PS) and the mental-work occupations (MA and TP) are plotted against time. Note that the "physical-to-mental" transition in the occupational domain of Figure 3b occurred at the same time that the "industrial-to-post-industrial" transition occurred in the sectoral domain of Figure 1.

The second major feature of Figure 3b is the extrapolation of the historical data to its projected terminus, the mid-point on the lower right boundary, representing 50% technical-professional and 50% managerial-administrative. This extrapolation of the fitted-curve shows that if the U.S. work force continues to evolve in the direction and at the rate of the past 80 years, its trajectory will carry it across the equator of the figure ca. 2010 and to its terminus at the (50%TP, 50%MA) point ca. 2080.

Figure 3d shows for each of the four occupational groups the extensions of their individual historical trajectories to the terminal values of their four-wise trajectory of Figure 3b, that is, PP = PS = 0, TP = MA =50%. This extrapolation projects two results: first, in less than a generation, that is ca. 2010, the work force will cross the point at which there are more workers in the technical-professional occupations than in physical-production; second, within the possible life expectancy of a female child born in 1990, that is ca. 2080, the work force will consist almost exclusively of workers in technical-professional and managerial-administrative occupations, there being only a vanishingly small fraction appearing in either of the current physical-production or physical-service occupational groups.

What Figure 3d clearly shows is that within the U.S., if the well-established trend continues, technical-professionals as a group, including para-professionals and technicians of all types,

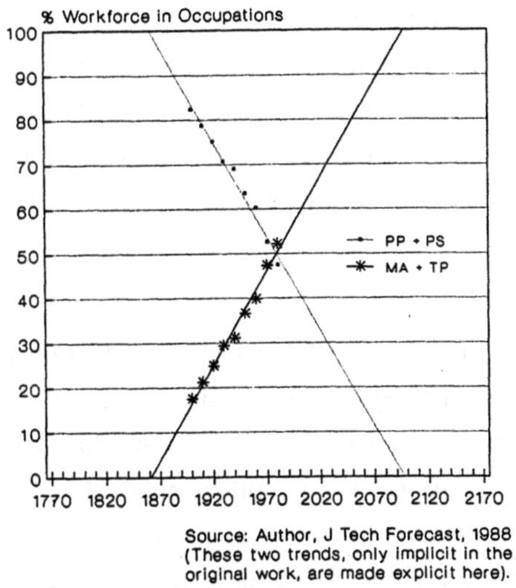

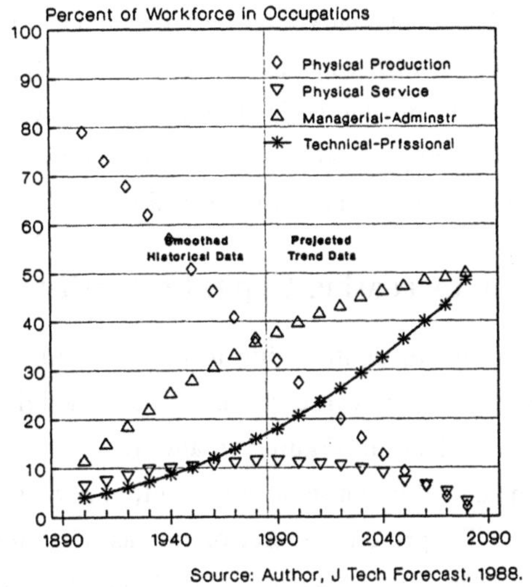

Figure 3c. Historical and projected evolution of US work force into white-collar (i.e., MA + TP) occupations

Figure 3d. Projected and historical evolution of US work force into technical-professional occupations

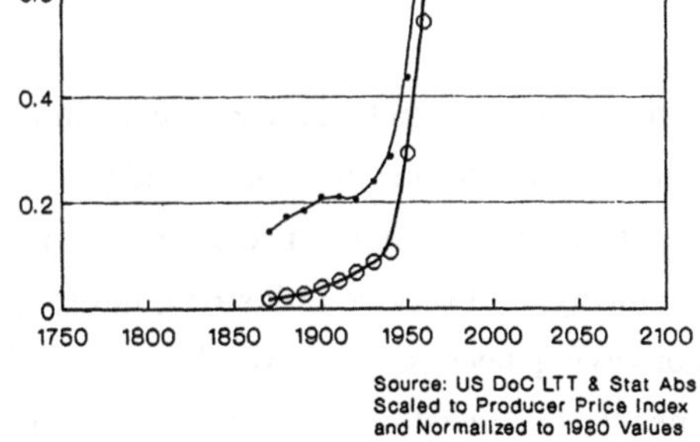

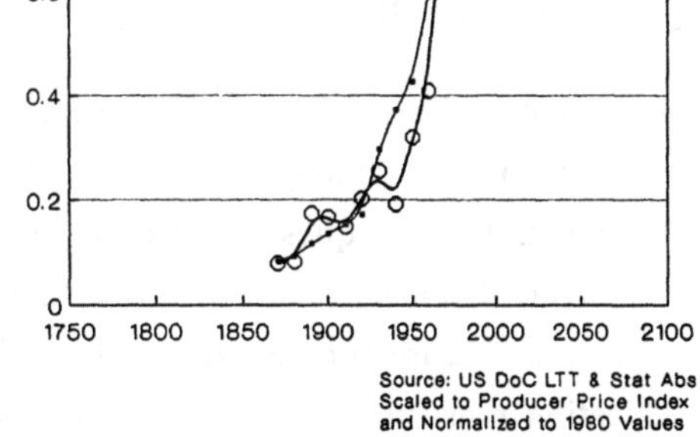

Figure 4a. Constant-dollar output per hour and capital equipment per worker in US agriculture

Figure 4b. Constant-dollar output per hour and capital equipment per worker in US manufacturing

will triple in relative size over the next two to three generations. The implications of these work force data for education in general and technology education in particular are clear: as U.S. shifts from blue-collar production to technical-professional service as the basis of its post-industrial economy, the U.S. educational system must shift gears to provide people the education they need to live and work effectively in that new "knowledge-based" economy.

The Increasing Capital Intensity of Modern Work

One factor contributing to this shift in the U.S. economy from labor-based production to knowledge-based service is the continued rise in the productivity of production, both agricultural and industrial. Figures 4a and 4b show respectively the dramatic increases in productivity, measured in constant-dollar output per worker, for agriculture and manufacturing [2].

The principal source of increases in long-term productivity growth is increased application of ever-more-advanced technology. Figures 4a and 4b also show the dramatically increased use of machines in agriculture and manufacturing respectively, as measured in dollar-value of equipment per worker. It is clear that equipment intensity, measured in terms of machines per worker, is the principal determinant of productivity, whether in manufacturing, agriculture, or service industries. Recent trends in industry also indicate that computer- and robotics-based machines are also the principal means to achieving otherwise unattainable levels of quality.

The lesson of increased effectiveness with increased capital per person has not been lost on U.S. service industries, which have for the last decade led the U.S. economy in capital spending. Figure 4c shows for the U.S. the level of capital spending for specific industries, the production and service sectors, and the economy as a whole. As indicated by the figure, the service sector increased its plant and equipment by almost a third, with business, legal, health, and financial services increasing by more than half.

Capital Intensity in Education

Given that the U.S. is moving into an era in which knowledge is a strategic resource, technical-professionals are strategic workers, and education -- as the conveyor of knowledge -- is a strategic industry, a question arises regarding how education is faring in use of technology.

Given also that, just as in agriculture, manufacturing, and service industries, in order to achieve necessary productivity and quality in conveying necessary knowledge and skills, education must use technology as a principal means, the question regarding how capital intensive education is becoming must be raised.

Figure 4d represents U.S. constant-dollar annual capital spending for K-12 education for the period 1890 to 1990 [2]. As the figure shows, for the first three-quarters of this last century, the trend in education's capital spending per child -- as in agriculture's capital equipment per farmworker and manufacturing's capital equipment per factory worker -- had been steadily upward. The figure also shows that between 1970 and 1985, capital spending per child suffered a decline of historically unprecedented depth and duration; for 15 years, capital spending in real dollars decreased by approximately one-third on a per child basis and more than a one-half on a per capita basis, to a meager $40 per child and $30 per capita respectively.

Given the point of view of this paper that the major improvements in the productivity of education -- as in agriculture, manufacturing, and service industries -- are attainable through greater and more effective use of machines, the recent history of capital spending in basic education -- including an unprecedented 15-year decline in capital investment -- should be an issue of major concern

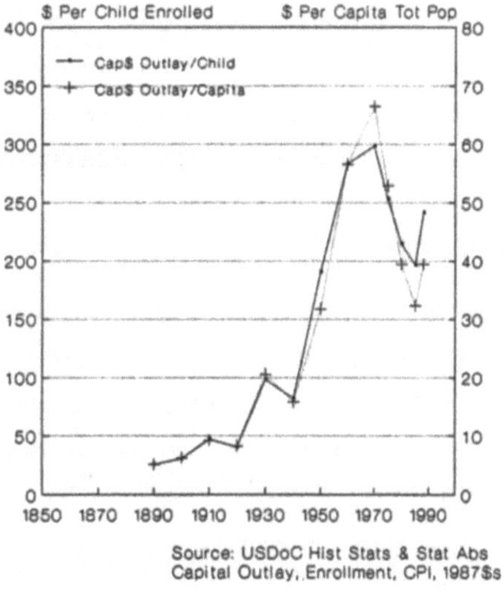

Figure 4c. Change in sector's plant and equipment including economy as a whole for the period of 1980 to 1987

Figure 4d. US constant-dollar annual capital expenditure in K-12 education showing major recession of 1970-1985

Conclusion

Behind the need to integrate advanced technology into technology education is a profound change in economic activity which is reshaping the work force of the industrialized nations in general and the U.S. in particular. The change in the work force is two-fold: first, there is the unprecedented decline in the relative size of manufacturing employment; and second, there is the unprecedented decline in relative dominance of manual-labor occupations. As this paper has shown, both changes are sharply demarcated, with 1975 being a watershed year.

Underlying the change in the manufacturing work force is the ever increasing use of productivity increasing technology. As this paper has shown, in manufacturing (and in agriculture), productivity measured in terms of constant-dollar value of output per worker and use of technology measured in terms of constant-dollar equipment per worker continue to growexponentially.

Given the nature of this new post-industrial economy, with its increasing numbers of technical-professionals and increasingly greater use of technology, the implications for education are clear: to meet corporate and government needs for well-prepared workers, education must become more technology intensive, both in terms of the technology it uses to teach and technology it teaches about. It is all so clear that for U.S. education to meet those needs by becoming more technology intensive, greater capital investment than has been made in the last two decades is required; this suggests that means to focus on capital spending in basic education, such as instituting capital equipment budgets, are in order.

References

1. Swyt, D.A. "The Workforce of U.S. Manufacturing in the Post-Industrial Era, *Technological Forecasting and Social Change,* 34, 231-251 (1988).
2. U.S. Bureau of Census. *Historical Statistics of the United States: Colonial Times to 1970, Bicentennial Edition,* Washington, DC: Oct. 1976 *and Statistical Abstracts of the United States,* Washington, DC: 1989.
3. U.S. Bureau of Census. *Long-Term Economic Growth of the U.S. Economy 1860-1965: A Statistical Compendium,* Washington, DC, Oct. 1966 and International Labor Office, *Yearbook of Labor Statistics.* Geneva, Switzerland: 1985.

New and Emerging Technologies that are Influencing Industry and Society

Jean-Jacques Duby

IBM Europe, France

Abstract: This paper attempts to predict the technologies that will be most influential over industry and society in the next decade. It covers in succession: mathematical technology, physics technology, chemical technology, materials technology, biotechnology, and information technology.

Key words: biotechnology, computer chemistry, information technology, materials technology, mathematical technology, physical technology

Introduction

The society we live in at the end of the 20th century has been shaped by four revolutions, at least: two agricultural revolutions and two industrial revolutions. The first agricultural revolution took place over the Middle Ages in Western Europe monasteries; the development of animal power and of crops rotation allowed a strong enough population growth to overcome, at long last, the devastating death sweeps caused by recurring epidemics or bad crops. The second agricultural revolution is just finishing. The generalization of agricultural automation and the improved selection of seeds have dramatically increased yields per acre as well as per capita, thus at the same time decreasing the farming population in developed countries to a few percent and creating food overproduction.

The first industrial revolution was introduced by the steam engine. This new source of power allowed the development of the first mass industries--steel and textile. It also made a new class appear in society--workers proletariat. The second industrial revolution was also a power revolution due to electricity. Electricity, easier to transport and use than steam power, further accelerated the development of industry and of a blue-collar class in society. But it can also be considered as the first technological revolution in the modern sense of technology, *i.e.*, trans-disciplinary and cross-industry. Indeed, electricity technology cross-bred with chemistry to produce electrochemistry, with metallurgy to produce electrometallurgy, with the transportation industry, etc.

NATO ASI Series, Vol. F 78
Integrating Advanced Technology into Technology Education
Edited by M. Hacker et al.

In recent decades, technology has taken the lead over industry. The technology revolution that we have been living, due mainly to the development of electronics and information technology, has had considerable impact on nearly all industries and on society, where it has considerably increased the tertiary sector and the white-collar population, and changed everyone's everyday life.

Are there other technological revolutions around the corner? Likely so -- but it may be risky to predict which technologies will be their main actors. The number one problem for the 19th century engineers was how to distribute steam power through shafts, pinions, and belts to each individual engine in the plant - and this problem disappeared overnight with electricity. The purpose of this paper is to present, among the main new or emerging technologies, those which, in the author's opinion, are most likely to profoundly influence industry and society on the way towards the 21st century. The reader may be surprised to see some "old" technologies in that list, like mechanics, for instance. The reason is that the author believes, rightly or wrongly, that these older technologies are being rejuvenated.

Also, this paper being presented at a Workshop on Technology Education, the author thinks that education in those older technologies may have been recently neglected, particularly in some European countries, and seizes this opportunity to express his views to international leaders in technology education. This lack of education in older technologies is even more serious due to the fact that modern technologies depend on older technologies to develop and improve the archetype of modern technology. VLSI electronics needs mechanics to position silicon wafers with submicronic precision, optics to draw complex circuitry, chemistry to etch the drawing into the material -- yet, mechanics, optics and chemistry are considered not really "in."

The outline of this paper will follow Auguste Comte's classification of sciences. This may look arbitrary, not to say nationalistic. However, this choice has another justification than mere convenience. It will try to reflect the growing interdependence between science and technology, hence between science and industry, science and society. The first agricultural and industrial revolutions were achieved without any scientific involvement, on purely empirical or engineering grounds. Watt's steam engine patent was granted in 1769 and Carnot's "Reflexions sur la Puissance Motrice du Feu" was published in 1824. Today, technology thrives on science, as is perhaps best exemplified by biotechnology and genetics. By contrast, a technology which has developed without solid scientific grounds may be less pervasive. Cars or planes could be glued together instead of welded or riveted, but engineers are reluctant to use adhesives until physico-chemists fully understand why glued parts stick together......or separate!

This paper will then start with mathematical science and technology, following with physics, chemistry, and biology. Topics unknown in the days of Auguste Comte will be added, namely materials and information sciences and technology.

Mathematical Science and Technology

The expression "mathematical technology" may come as a surprise. What is meant by this is that the combination of applied mathematics and computers is turning into an emerging technology. Mathematical engineers are being called upon to solve technical problems -- optimization, estimation, computation by all disciplines and technologies -- the same way a mechanics engineer may call an electrician to automate a device or a chemist may consult with a fluid dynamics specialist to understand convections in a reactor.

Two areas in applied mathematics will have a growing importance in the next decade. First, modelization and simulation will be used more and more widely to replace experimentation. Already today, "digital" wind-tunnels replace the real ones for the vast majority of preliminary tests, leaving only the last final stages of the prototype to be tested in real. This is due of course to the improvements in computer speed, but even more in mathematical techniques like finite elements or Monte-Carlo methods.

Second, statistics will have a growing importance in every industry. As the quality requirements get stronger, production controllers will need robust sampling and estimating techniques, whether they manufacture yogurt or car speedometers. Furthermore, the development of life sciences, where experimentation is more difficult and scarce -- *e.g.*, in medicine or in agriculture -- will need further improvements and wider dissemination of recent sophisticated statistical techniques, like bootstrap or jackknife.

But applied mathematics will not be the only part of mathematics used to solve industrial problems. More and more, solutions to very difficult industry generated problems will require advanced pure mathematics. For instance, it is rarely acknowledged that sound encoding on compact disks uses Galois theory, or that robotics engineers are contemplating using algebraic geometry to detect moving pieces collisions.

From an educational point of view, this double trend has consequences on the way mathematics is taught for the majority of youngsters. Mathematics should be taught as a service discipline more than as a science. In other words, it is more important to teach the use of mathematical tool kits than the demonstration of theorems. However, some insight into profound areas of mathematical research should be useful, not only for general culture, but also because, however abstract, these areas will one day be applied to resolve an until then unsolvable industry problem. As for the minority of pure mathematicians, they should learn that what they do could be used.

Physical Science and Technology

With the very fast pace of evolution of technologies, there are several areas of fundamental physics which are directly involved in industrial technology development. Statistical physics' treatment of order and disorder plays a direct role in complex fluid dynamics and in understanding properties of some new materials. In the study of matter, light interaction is paving the way for a pervasive use of lasers in mechanical engineering (cutting, drilling, surface processing, etc.) and in biomedical applications (microsurgery). A better theoretical understanding of surfaces and interfaces will be required as the size of VLSI circuits gets dangerously close to a few atomic length.

From a technological point of view, electronics will continue to be the most influencing technology. Energy technology will also play an important role, although for the mid-term at least, one can fear it will be part of the problem more than part of the solution. And the oldest physical technology, mechanics, will find a second birth.

In electronics, miniaturization of silicon will go on for the next 10 to 20 years. Gallium arsenide and other III - V components will develop their niche in optoelectronics, although it is unlikely that they will displace silicon. Most of this evolution will be by incremental steps -- improvements in manufacturing precision, in design tools, particularly in testing devices and methodology. However, one important technological breakthrough will be needed as the circuit geometry reaches one third of a micron: the replacement of visible light or even ultraviolet light optics by X-ray optics. As there are no classical optic devices for X-ray, this new technology will have to be built from scratch, first in the research and development laboratory, then in the plant. Another emerging technology in electronics is also related to miniaturization, as VLSI circuit behavior steps into quantum mechanics domain: quantum wells and similar devices will likely influence electronics and optoelectronics. It should be noted that innovations in electronics will not come only from the high-end sectors like computers or space or military applications. More and more will come from consumer electronics, as was the case already for compact disks, digital audio tapes, and liquid crystal displays.

Power electronics does not enjoy the same fame as low current electronics. Yet, it will play a growing part in transportation and power industries as requirements for less pollution and fossil fuel conservation decreases the use of oil. Already today, the French TGV (high-speed train) pulling engines are full of thyristors -- the engines themselves occupying but a small volume beneath the floor.

Energy technologies will be increasingly called upon as the energy shortage and the pollution issues both get more acute. Unfortunately, no major breakthrough can be hoped for in the near or mid-term future. Controlled nuclear fusion is still decades away. Solar energy is still expecting miracle photovoltaic materials. Electricity storage is also waiting for lightweight batteries with high energy density. It may be sad but more reasonable to recognize that the most important energy technologies in the coming years will be involved with nuclear waste recycling. To be a little more optimistic, improved thermodynamics, fluid dynamics, and chemistry may allow for a really clean internal combustion engine without the help of the catalytic exhaust fix.

Mechanics, maybe because it is the oldest physical technology, has been too often despised in engineering education. Yet, it recently sent severe "forget me not" messages. The Shuttle explosion was due to the mechanical failure of a gasket. The DC-10 crash near Paris was due to a bent cargo door latch. Because they are old, and also probably because they are cheap, mechanics technologies are taken for granted and not given enough attention. Nevertheless, precise mechanical engineering, carefully designed structures and moving parts are necessary to almost every other technology, be it in the production process or in the product itself. The example of advanced consumer products like camcorders is illuminating. Sure a camcorder is full of electronics, but it requires both advanced "tape automated bonding" techniques to manufacture and sophisticated mechanics to control the magnetic tape on the reels. Besides "classical" technologies, which are sometimes pushed to the limits (as illustrated by the scanning tunnel microscope that "senses" individual atoms), mechanics recently introduced a variety of new technologies like laser cutting or drilling, plasma or electron beam welding, ultra high -speed quenching, etc. Often combined with electronics in "mechatronics," mechanics will be just as pervasive.

Chemistry Science and Technology

Today, chemistry enjoys as bad press as mechanics. The latter gives greasy hands, the former pollutes. Yet, from a scientific point of view, chemistry is progressing rapidly towards understanding the innermost mechanism of chemical reaction, down to the atomic level. Advances in quantum chemistry as well as in mathematical technologies mentioned above already allow *ab initio* computation of relatively complex molecules (tens of atoms).

From a technological point of view, those progresses allow the emergence of what could be called "computer chemistry." This technology will provide a whole set of computer tools for chemical process design: reaction databases, reaction modelling, molecular modelling, molecular engineering. Combined with advanced artificial intelligence techniques like constraint logic programming and non-shallow reasoning, new reaction models will allow automated control of reactions. Molecular engineering will help design new materials and new drugs (see below). Computer chemistry should also revolutionize the teaching of chemistry.

One particular chemical technique should be drastically changed as a result of this evolution. Catalysis should leave its present empirical stage to reach "programmed" catalysis based on artificial materials designed for specific reactions. Three-dimensional catalysis, like the one taking place in zeolites, could be one of the most promising emerging catalysis technologies.

Materials Science and Technology

From a scientific point of view, physico-chemistry is making considerable progress towards the understanding of the links between the microscopic structure of a material, down to the atomic level, and its macroscopic properties, like mechanical, electromagnetic, or surface characteristics.

These scientific advances, combined with new technologies like "computer chemistry" or new mechanical processing techniques will give birth to an outburst of new materials. In this process, the battle between metallic and organic materials will be intensified. Metallurgy will rely on new technologies like thermal treatment (directional solidification, ultrafast quenching), surface treatment (laser, plasma) or sophisticated alloys. Organic materials will rely on new computer chemistry technologies. Composite material technology will also develop, although it presently lacks solid scientific background. All these new technologies should allow the design of a material to match specific requirements of weight, strength, surface, resistance, even anisotropy, for instance.

Two specific materials technology could be soon emerging: metallo-organic compounds and disposable materials. The former could yield new materials with metallic electromagnetic properties and organic mechanical properties like conductive polymers. The latter could provide an answer to the ever growing problem of waste recycling or disposal: the designer of a new material could take into account what to do with it after its useful lifespan -- moving from planned obsolescence to planned senescence!

Life Sciences and Biotechnologies

Biology may be the science which can be expected to make the most considerable breakthroughs in the coming decade: genetics with the deciphering of the DNA code, biochemistry with the understanding of the cell as the elementary biochemical reactor, immunology with the understanding of the different identifier mechanisms of cells and viruses.

These scientific breakthroughs will in turn accelerate the development of biotechnologies, and principally of genetic and enzymatic engineering. Applications will be innumerable, from the production of specific proteins or amino acids for medical or food applications to the engineering of new bacteria for waste disposal or ore lixiviation. Other biotechnologies should have considerable impact on medicine, like monoclonal antibodies which are already commonly used for diagnosis purposes and should be used extensively to design "targeted" drugs releasing their treating chemical exactly where it is needed and not throughout the whole body.

A new biotechnology might be emerging to help both with environment and agricultural surplus problems, leading to innovative--*i.e.*, non-food related--uses of biomass. Several attempts have already been made in the direction of alternate energy sources, but more promising still are trials like the use of corn sugar in PVC to make plastic more biodegradable.

Information Science and Technologies

We will consider in this section informatics, telecommunications, and automation.

In **informatics**, computer science should make definite progress towards alternate hardware architectures (*i.e.*, non-Von Neumann), and some advances in the understanding of software semantics.

From a technology point of view, hardware technology will be changed considerably by the emergence of new, non-Von Neumann architectures, hyperparallel architectures, fault tolerant architectures for certain hyperconnective architectures, data flow architectures, even neuronal architectures possibly. Unfortunately, it is likely that software technology will not evolve at the same pace. It will certainly make incremental progresses, particularly in the areas of parallel or cooperative processing to cope with new hardware architectures and of software engineering, *e.g.*, for "reuse" of preexisting code, but no really new software technology seems to be appearing -- unless one considers as "new" technologies logic or object oriented programming, which should be more and more widely used.

In **telecommunications**, there is no specific scientific discipline to support the technology, but different disciplines may be called upon to solve technological problems, like mathematics to provide efficient error correcting codes or solid state physics to produce new optoelectronic devices.

The new pervasive technologies in telecommunication will all be focused on high bandwidth. This will require mastering ultra high frequencies (in the gigaherz range) and designing new protocols to control the flow of enormous amounts of data (in tens of megabit/sec). No new media technology is to be expected, although ether and optical fibers will be more used and copper less -- but those are not new technologies, only their pervasive use at very high bandwidth and frequency is.

In **automation**, the technology of the working station (numerical command machine tool, robots, flexible assembly cells) is now mature. The new emerging automation technology is now aiming at networking those working stations into a production network (computer assisted production control), this network itself being embedded into an enterprise network comprising development, sales, buying, billing, etc. (computer integrated manufacturing or CIM). The enterprise network itself will be extended upstream (to suppliers and/or subcontractors) and downstream (to distributors and/or even customers) to produce the so-called "extended enterprise." CIM will undoubtedly be the new automation technology for the coming decade. It will speed up response to market demands, shorten the production cycle, reduce inventories and, therefore, costs. Even small and medium enterprises will need a CIM system as large buyers will place orders through their own.

Conclusion

In such a limited space and time, this paper could by no means be an exhaustive listing of all the new technologies that will influence industry and society. The author only tried to list what he views as the wide areas of technology that are more likely to play a major role in the coming decade. It will, of course, be interesting to compute his hit ratio in year 2000. In the meantime, the author would like to conclude with a few messages that, as an industry executive in charge of science and technology, he would like to send to educators in charge of technology education:

- Modern technology depends on science. Both the technology teacher and the science teacher should consistently recall this dependence, the former by referring to each technology's scientific background, the latter by showing applications of his/her fundamental discipline.
- Modern technology, like modern science, is trans-disciplinary: technology education should aim at a multi-disciplinary technical culture, however difficult it may be in some cases (*e.g.*, mechanics and biotechnology).
- So-called "advanced technologies" need "not advanced technologies": technology education should not be restricted to "fashionable" technologies.

Technology education is getting more and more important as technology is becoming the primary development factor in industry and as the control of the market is more and more dependent on technology control. For today's industrial countries, efficient, up-to-date, pervasive technology education is a prerequisite for the future of their industries and their people.

II. Contemporary Technology Education

Not only the concept of "advanced technology" in the title of the conference needs to be defined but also the concept of "technology education." To be able to discuss the integration of advanced technology into technology education, one should have an impression about the state of affairs in contemporary technology education. This is relevant to determine how and where advanced technology will fit in the total field of technology education.

Dr. Ernest Savage (Bowling Green University, USA) presented a conceptual model for technology education, in which the way development of new products in technology takes place was shown. The model represents technology as a dynamic process. In such a dynamic model there is a place for the integration of new and advanced technologies.

Mr. Robert McCormick (Open University, UK) gave a survey of developments in the past that had led to the current practice of technology education. Various rationales for teaching technology as a part of the education of all children were presented, as well as various traditions out of which technology education has grown.

In his presentation **Dr. Ronald Todd** (Drexel University, USA) developed a philosophical basis for contemporary technology education. This basis included the concepts of products, processes, procedures, and knowledge. He made clear that there is an essential relationship with science and stressed the importance of developing a common language of technology.

Dr. Barry Stern (Department of Education, USA) underlined the relevance of technology education as a component of education for all students nationwide. He pointed out, however, that curricular reform will have little impact without structural reform of the education system.

Discussions focused on the various goals of technology education that are regarded as important nowadays: students should not only acquire knowledge and skills to use technological products, but also learn to contribute to a technological world, evaluate technology, and act as critical citizens. Learning of problem-solving skills is generally seen as important in technology education. Technology education should also function as an introduction to an important field of possible future careers for students. These goals make technology education an important element of all students' education, although it is not always recognized as such. Therefore participants felt a need to state a number of recommendations to NATO countries which call for the stimulation of a strong technology education practice. These recommendations were formulated during the conference and they have been included in the Proceedings.

Determinants of Advanced Technological Content in Technology Education Curriculum

Ernest N. Savage

Bowling Green State University, USA

Abstract: Technology is defined and its characterisitics are identified. The technological method, its components, and its relationship to other disciplines are addressed. Recommendations for advanced technology education content determination are presented. It is concluded that advanced technology curriculum should focus on the entire technological method rather than just its knowledge organizers and technological process.

Key words: technological method, bio-related technology, communication technology, physical technology, technological process, characteristics of technology

Introduction

In order to determine curriculum content for technology education, it is necessary to consider the following questions:

1. What is technological method and how does it relate to methods in other disciplines (science, humanities/arts, social sciences)?
2. How do the components of the technological method allow for the identification of technological content?
3. How can the technological method be used as a content model for technology education?

Definition of Technology

Technology is a body of knowledge and the application of resources using a systematic approach to produce outcomes in response to human needs and wants. [8]

Characteristics of Technology

The nature of technology and technological knowledge can be characteristically examined as object, process, knowledge, and volition [5]. Technology as object or artifact relates to the physical output (tools, consumer products, etc.) of a material-producing activity. These outputs or objects are essential to everyone in our culture. For example, television can take students to

NATO ASI Series, Vol. F 78
Integrating Advanced Technology into Technology Education
Edited by M. Hacker et al.

the moon and to the bottom of the sea. Calculators can solve problems faster than the human brain. And computers can instantly retrieve millions of information bits. Word processors can help students write and edit [1].

Technology as process involves the making and using of the objects of technology. Frey [2] identifies three structures in which technology may be considered a process: (a) as a material-object transformation process -- without affecting the form (heat treating) or material transformation (steel making); (b) as a design process -- an array of systematic methods (mathematical and physical modeling, dimensional analysis, testing methods, systems engineering, and "parameter variations") [3]; and (c) as a decision-rules process -- made within a social context and based on many non-technological factors.

Three features of technology as knowledge are (a) its basis in praxis (efficient practice), (b) its aims and purposes, and (c) its level of complexity [2]. Praxis relates to the importance of practice in developing technological knowledge in order to produce a pragmatically oriented (doable) production process. The aim or purpose of technological activity is to put into concrete form ("Does it work?") the ideas of the engineer or the applied scientist ("Is it true?"). Skolimowski [9] argues that the knowledge generated from technological ventures results from the quest for effectiveness based in efficient action (praxis). The level of complexity of technological knowledge ranges from the lowest level of artisan skills through more complex levels of technical maxims, descriptive laws, and technological theories.

Technology as volition incorporates aims, intentions, desires, and choices [5]. Persons who produce technological objects must make conscious decisions regarding design and process factors. Users of the technological object must address social, political, ethical, and economic issues.

Relationship to Other Disciplines

Bodies of knowledge are often grouped into broad categories according to the nature of content and methods of knowing and doing work. In this respect, each body of knowledge (science, technology, social science, humanities, arts) occupies a unique and equivalent position.

Furthermore, bodies of knowledge are interrelated (see Figure 1). Each contributes methods and content useful to others. For example, human values (humanities) and aesthetics (arts) facilitate and constrain technology. Scientific theory and practice (science) contribute to technological methodology. Social and personal behavior patterns (social science) serve as parameters for technological activities. Communities of technologists (technology) actively incorporate this knowledge in solving problems. Bodies of knowledge have distinguishing features, as represented in Figure 2.

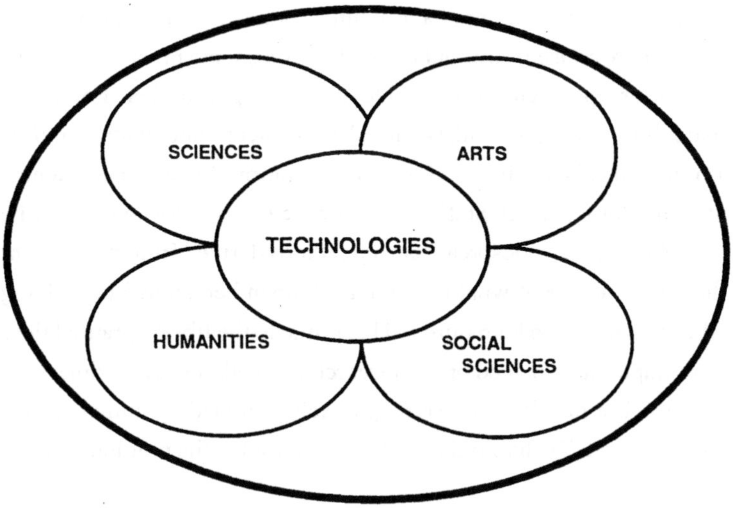

Figure 1. Relationships of bodies of knowledge

	TECHNOLOGY	SCIENCE	HUMANITIES/ ARTS	SOCIAL SCIENCES
DEFINITION	STUDY OF THE HUMAN MADE WORLD	STUDY OF THE NATURAL WORLD	STUDY OF VALUE	STUDY OF HUMAN BEHAVIORS
METHOD OF INQUIRY	TECHNOLOGICAL METHOD	SCIENTIFIC METHOD	ECLECTIC METHOD	SCIENTIFIC PHENOMONOLOGICAL CRITICAL SCIENCE METHODS
CONTENT EXAMPLES	BIO RELATED COMMUNICATION PRODUCTION TRANSPORTATION ETC	BIOLOGY CHEMISTRY PHYSICS MATH ETC	HISTORY PHILOSOPHY LANGUAGE ETC	SOCIOLOGY ANTHROPOLOGY PSYCHOLOGY ETC

Figure 2. Features of bodies of knowledge

Communities of practitioners have unique methods of generating new knowledge. Although there is more consensus about methods among some practitioners than others, each has methods which are peculiar to its problems and opportunities. For example, the primary research paradigm among scientists includes problem statements, careful observation, the statement of hypothesis, testing, and theory generation. Most social scientists share the same positionist methodological orientation as natural scientists. However, there is still considerable flexibility, including the widespread use of methods derived from phenomenology and critical science, due to the nature of what is studied -- human behavior. Methodological flexibility is probably the greatest in the humanities. This is due to the liberty granted the individual as well as the greater importance placed on the researcher's methods. The primary method of knowing for technologists is a problem-centered approach governed by human needs and wants called technological method. Technologists' activity is based in the pragmatic desire or need to solve human problems.

The Technological Method

Human needs and wants lead to the identification of problems and opportunities. These problems/opportunities are then addressed by applying resources, technological knowledge, and technological processes to reach an evaluable solution that has impacts. This interaction is known as the technological method (see Figure 3) and is the essential model for "doing" technology. This method can be used at the personal, organizational, institutional, or global level. Therefore, it identifies a means for engaging in individual or collective technological action [8].

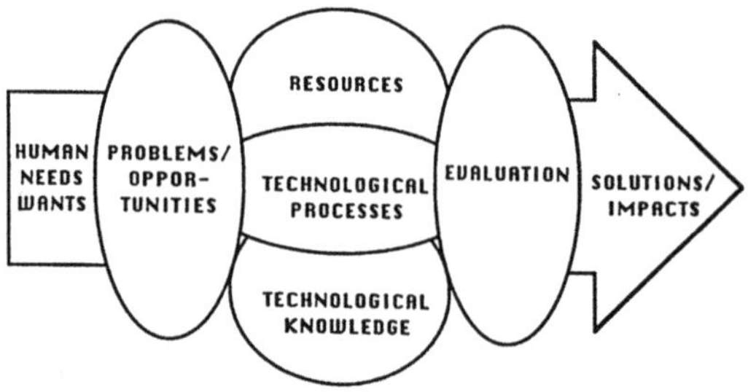

HUMANS: Needs Wants	**TASKS:** Problems Opportunities	**RESOURCES:** Personnel Tools/Machines Data Materials	 Energy Capital Time	**TECHNOLOGICAL PROCESSES:** Analyzing Realizing Testing
TECHNOLOGY KNOWLEDGE: Bio-Related Technology Communication Technology Physical Technology		**EVALUATION:** Satisfactory Exceptional Unacceptable		**ACCOMPLISHMENTS:** Solutions Impacts

Figure 3. Technological method

Human Needs and Wants

Humans are the focal point of all that we do in technology. Therefore, the human dimension must be reviewed over a lifetime and consider all the roles a person plays and all the ways a person will need to continue learning. For example, we may play roles such as student, worker, citizen, parent, family member, entrepreneur, investor -- the list goes on. In all these roles we are faced with problems and opportunities, both of which we need to deal with systematically. In doing so, we learn, apply, and add to what we know and how we use and expand our knowledge base. All of us need to be technologically competent to a level that enables us to function. This may be at a very high level of technological ability if our profession requires such a level of application.

As people develop from childhood to adulthood, they interact with technology at varying levels and for different reasons. As a result, levels of competence are expanded, but in different ways. For the young child, learning is often informal. During adolescence, learning occurs in a formal, structured environment. This is where technology is studied in an organized process. As we reach adulthood, schools, colleges, and universities provide theory and practice needed to meet professional work responsibility. In fact, at this level individuals often contribute to the overall technological body of knowledge.

Problems/Opportunities

Problems are questions or matters involving doubt, uncertainty, or difficulty. They deal with choices of action which might ameliorate unsatisfied human needs and wants. Opportunities, on the other hand, can be planned and, in fact, created.

Problems facing humans involve personal, social, and environmental issues. Solutions to these problems may utilize technological, social (including political), and educational (leading to behavior modification) means (see Table 1). While problems tend to be "now" oriented, opportunities are future oriented. The future can be studied, planned, and anticipated. Just as there are problems of today, there are scenarios of an optimum future.

Types of Problems	Technological	Social	Educational
Personal (e.g. Loss of Vision)	Contact Lenses	Affordable Medical Care	Effect of Looking Directly at Sun
Social (e.g. Food for the Masses)	Chemical Herbicides and Pesticides	World Hunger Crusade	Awareness of Situation in Other Nations
Environmental (e.g. Clean Energy)	Solar Panel	Legislation (i.e. Clean Air Act)	Need to Recycle

Table 1. Types of solutions

Resources

Resources are those items needed to accomplish technological ends. They are personnel, tools/ machines, data, materials, energy, capital, and time. The use of these resources must be optimized to ensure efficiency and renewability where appropriate. Constraints must be considered including physical limitations and regulation.

Personnel

The problems, needs, and opportunities perceived by people stimulate the development of technology. People design and create technology by combining previous knowledge with new ideas. People make policy decisions that promote or constrain technological research, development, and growth. People also provide the labor upon which some industries depend. Finally people are consumers and users of the products and services of technology. People, using tools and machines, direct and control the processing of materials, energy, and information as the basic resources of technology.

Tools/Machines

Tools, including hand tools and machines, provide the means for humans to process and change the basic resources of materials, energy and information. Tools are used by people to reach desired goals or outputs such as different and desired forms of materials (a product or structure), energy (light or heat), or information (printed words or electronic signals). As we seek to employ tools in the technological process, we must consider constraints such as availability, access, and whether suitable tools have yet been invented to undertake the desired processes.

Data

Technology has developed rapidly because of the information explosion. The first information explosion, represented by the production of books using the printing press, laid the foundation for the technological upheaval that started in the Middle Ages and culminated in the Industrial Revolution. Information in this form was largely symbolic and static. During the past few decades there has been a new information explosion resulting in a doubling of technological data every five years. Data have taken on more dynamic forms as signals that can be transmitted electronically for purposes of communication or control. Management of an every-increasing database must be controlled.

Materials

Materials represent the physical stuff from which technological devices, structures, and systems are made. Materials come in natural and synthetic forms. Synthetic or manufactured materials far outnumber the relatively few raw materials that serve as natural resources for the finished products and artifacts of technology. Raw materials can be classified as renewable or non-renewable. Synthetic materials can be developed (engineered) to fit specific requirements and provide improved properties and characteristics. As we are constrained by our limited resources, we must ensure that the development of new synthetic products is controlled.

Energy

Energy is used to drive tools and machines, to process materials into manufactured products, and to create structures. Energy can be generated from selected non-renewable raw materials such as coal, oil, and gas. Energy can also be converted back and forth into different forms such as chemical, electrical, thermal, mechanical, light, and sound in response to the demands of an identified problem. Energy forms found naturally, such as solar, gravitational, and geothermal, provide the capability for renewable energy. Proper management of energy resources is essential to world market competition success.

Capital

Any form of wealth (e.g., shares of stocks, cash, land, building, equipment) is a capital resource.

Time

In the prehistoric era, time was measured in periods based upon the rising and setting of the sun. In the agricultural era, time was measured in days and seasons. In the industrial era, time was measured in terms of hours, minutes, and seconds. In the information era, time is measured in terms of smaller and smaller fractions of a second. Time represents an essential resource when all the other resources are brought together to support a process. Processes, as sequences of actions and change that yield results, require time -- no time, no process; too much time, low efficiency.

Technological Knowledge

The body of knowledge for technology must be identified based upon content that is unique to the study of the human-made world. The organizers must consider living and non-living content, the shaping of those entities into useful products, and the ability for "end-users" to access those products. As such, technological knowledge can be classified as bio-related technology, communication technology, and physical technology.

Systematic Representation of Technological Knowledge

Technological knowledge can be represented systematically using the basic systems model. The elements are input, process, and output (see Figure 4).

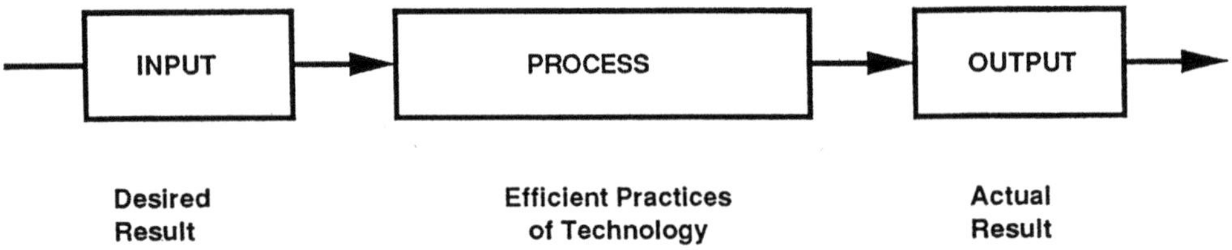

Figure 4. General systems model

The input is the command given the system. The input command is the desired end or the result the system is expected to achieve.

The efficient practices of technology are the processes identified in the general systems model. The "process" portion of the model provides the boundaries of the unique discipline of technology. Within the process boundaries are to be found the concepts, principles, generalizations, and unifying themes of technology. This qualifies the field of study to be considered as "process education." The inputs and outputs link technology to other systems; they are shared with other disciplines. While understanding of the inputs and outputs of technological systems is necessary, some of this knowledge is the responsibility of other disciplines. For example, the explanation of the chemical composition of different plastics is the responsibility of physical science. However, technologists need to know the physical characteristics in order to select the most desirable material and choose appropriate forming practices.

The output of the system includes the ends or goals (products, services desired, or undesired results) as they become parts of the newly accommodated environment within the sociocultural context. Each technology system has a unique set of processes that accompany appropriate inputs and outputs. Further explanation of each system follows.

Bio-related Technology Systems

Bio-related technology applies biological organisms to make or modify products. The system includes the production process techniques relating to the products of agriculture and medical technology, and the biological processes related to fuel and material production. Proper management of these techniques can result in a higher quality of life for all members of our global society. Mismanagement or misunderstanding can result in serious damage to ourselves and our environment.

The process techniques (see Figure 5) of the bio-related technology system are as follows:

Propagating To develop a living entity. The implications in the health and agriculture fields are obvious with respect to humans and animals. Fermentation and protein applications in foods, beverages, and materials also apply.

Growing To develop, or increase in size, living things by synthesis, intake, or manufacture. Humans, plants, and animals are all "growth monitored." Crystal manufacture for the electronic industry also applies.

Harvesting To gather and store entities. In the clinical sense, this technique relates to "giving birth" in human and animal entities, as well as accumulating products of agricultural and bio-process manufacture.

Adapting To adjust to a change in the environment. This is the essence of technology: to adapt to our natural environment. Through processes of protection, enhancement, management, and development, people in health care and services, agriculture, food and material production, waste management, and regulation have enabled and, at times, directed our adaptation to our environment.

Treating To apply a specific procedure to cure or correct a disease or pathological condition. When the human, plant, or animal output is improper, the necessary corrective measures must be taken. This is also true for materials in cases such as oil spills.

Converting To change into something of different form or property. Experiments across species, such as combining potatoes with algae to produce a more nutritious product, are common today [7].

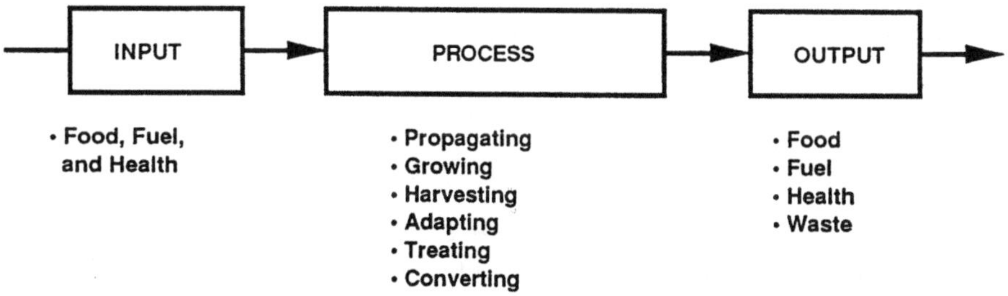

Figure 5. Bio-related technology systems model [7]

Communication Technology Systems

Communication technology focuses on the processes and techniques of encoding, transmitting, receiving, storing, retrieving, and decoding graphic and electronic messages. Our global society is interconnected with communication networks that are superior to any communication system in the past. Systems of the past and present, as well as models of futuristic communication systems, should be studied to allow students to develop an understanding of how they work, why we depend on them, and how they can be improved.

The following are the process techniques of communication technology systems (see Figure 6):

Encoding	To recode or modify information into a desired format or pattern for a specific method of transmission or storage.
Transmitting	To convey information from one location to another.
Receiving	To recognize and accept information that has been transmitted.
Storing	To record and file information for use at a later time.
Retrieving	To recall information from a stored condition.
Decoding	To convert recoded or modified information into an acceptable format for use [4].

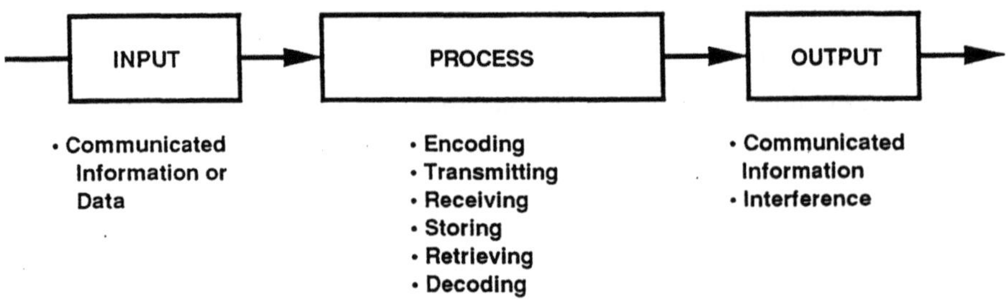

Figure 6. Communication technology systems model [4]

Physical Technology Systems

The physical technology systems reflect technology that involves the construction and manufacturing of goods and materials as well as the transportation of people and products. As in all technological endeavors, the utilization of energy and power is essential to efficient action. The physical technology systems model is depicted in Figure 7. The technical component of these systems can be classified as construction, manufacturing, transportation, and energy/power utilization [7].

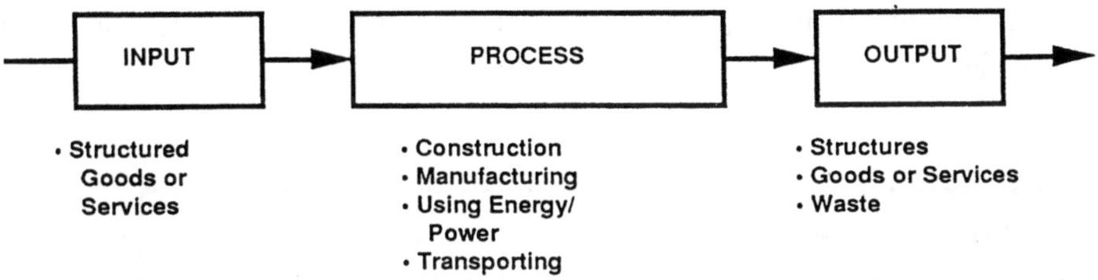

Figure 7. Physical technology systems model

Construction Technology

Construction technology employs technological practices to build and service structures on a site. Construction is a major industry in contemporary society and it has dramatic influence on the environment in which people live. For these reasons it is necessary for all people living in technologically advanced societies to obtain fundamental understanding of construction practices. Figure 8 presents the relationship among the construction technology system inputs, processes, and outputs.

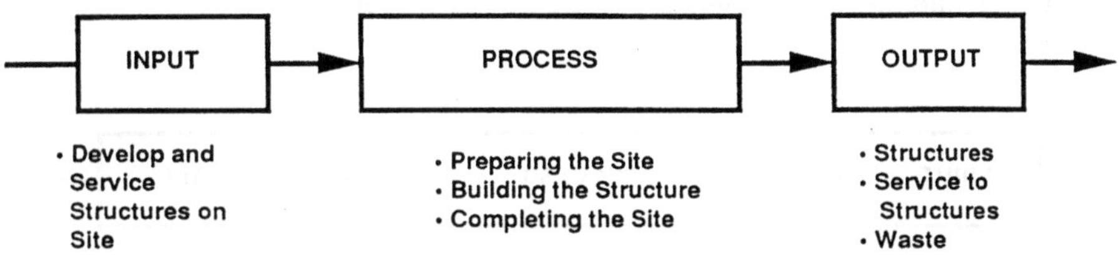

Figure 8. Relationship among construction systems inputs, processes, outputs

Manufacturing Technology

Manufacturing technology involves the production of goods within a facility. The manufacture or service of a product could be viewed in the same manner as the conducting of an orchestra. Each component of manufacturing, like each instrument of the orchestra, may be isolated for study. Only after individual concepts are brought together can a full understanding or appreciation of the musical product or the manufacturing technology system be obtained.

Figure 9 represents the relationship among the manufacturing technology system inputs, processes, and outputs.

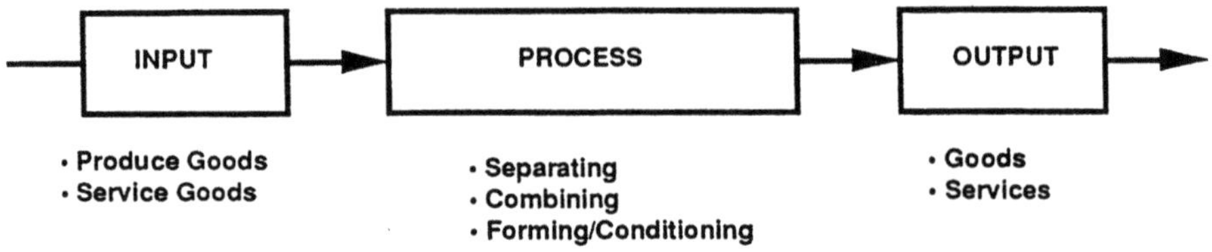

Figure 9. Relationship among manufacturing systems inputs, processes, outputs

Transportation Technology

Transportation technology is used to move people and products within a society. Transportation technology addresses four domains: terrestrial, marine, atmosphere, and space. Transportation options in a society can create opportunities for technological and cultural growth by enhancing mobility and providing product accessibility. Figure 10 represents the relationship among the transportation technology system inputs, processes, and outputs.

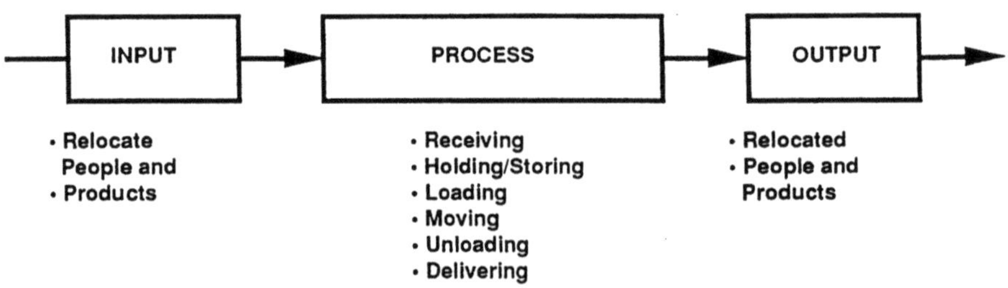

Figure 10. Relationship among transportation systems inputs, processes, outputs

Energy/Power Technology

Energy/power technology focuses on the practices utilized to develop, install, and maintain machines and other devices to support all of the fields of technology. Because energy utilization and power application are fundamental to the production of most products and the operation of most enterprises, it is imperative that students attain a basic understanding of the processes and principles underlying this system. Figure 11 represents the relationship among the energy/ power technology system inputs, processes, and outputs.

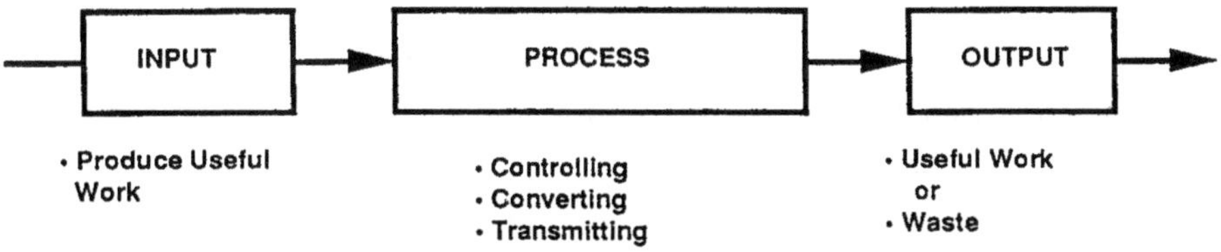

Figure 11. Relationship among energy/power systems inputs, processes, outputs

Technological Processes

Technological activity is not random. It must be systematized in order to function efficiently and effectively. A system is a definable relationship of the parts to the whole. Technological processes (Figure 12) provide the interface between resources and technological knowledge in order to solve problems or exploit opportunities. The heart of the system processes is the ability to analyze, realize, and test.

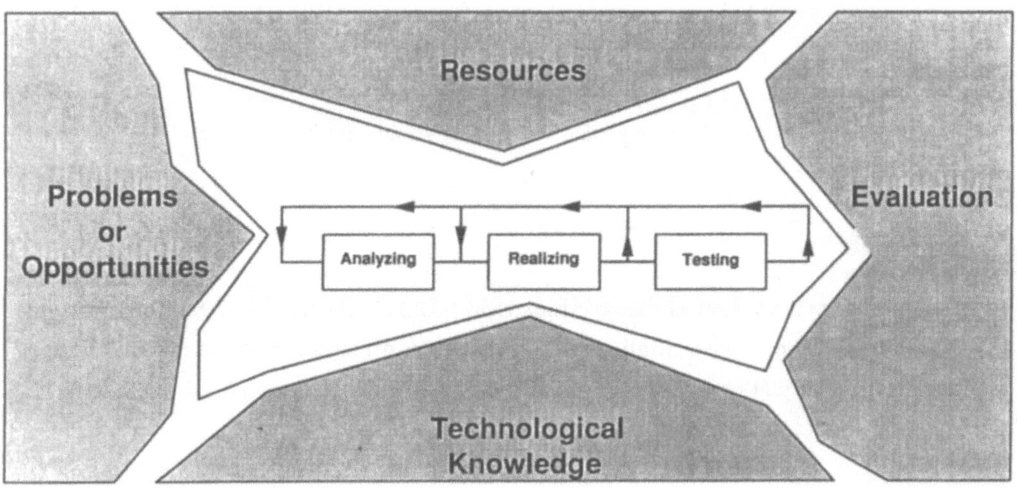

Figure 12. Technological processes

Analyzing

Analysis must occur to determine the nature of a problem or opportunity. It is usually operationalized through the development of a design brief which describes what the solution to the problem should do and what constraints are being imposed to keep the process on track. It also includes the actions of information gathering and the generation of alternative solutions.
Information gathering Familiarization with the factors which influence the problem, such as past history and unique characteristics.
Generation of alternative solutions Use of creativity and such processes as brainstorming to generate many alternatives to an initial solution.

Realizing

Realization includes those actions necessary to turn an idea into reality, the selection of a most appropriate solution, and the development of that solution into a prototype.
Selection of a most appropriate solution Based on the demands of the problem and the design brief, a "best" solution is selected.
Developmental work Reality and practicability are sometimes difficult to reach in the development of a solution. Through the generation of drawings, evolutionary models, and other 2-D and 3-D devices, theory can be brought into practice.
Prototype development Prototypes should be operational, if appropriate, and demonstrate a level of quality to communicate the intent of the finished product.

Testing

Testing determines if a product or process works. It should be based on specific criteria or objectives.

Feedback

Each component of the systems process is inextricably linked; the results of testing may require modification of the developed solution, which could also affect the analysis stage. Therefore, feedback is necessary at every stage.

Evaluation

The output of the technological method must be evaluated against the requirements established in the original problem statement to determine if the outcome is satisfactory, exceptional, or if it is unacceptable in part or totally. Unacceptability would require reworking and retesting until the criteria established in the design brief and original problem statement are satisfied.

Solutions and Impacts

The technological environment is a product of resources and technological processes. To study resources without technological impacts would present an unbalanced view of technology. Therefore, as is done with the resources, technological impacts must be considered as each unit is studied. In the personal arena, such areas as values, ethics, norms, and psychological (including techno-fears and techno-phobia) and physical influences should be studied and analyzed for decision making and problem solving. Environmentally, presentation of the impacts of technology should be sensitive to the ecological systems of water, space, land, and air. Socially, technology also affects the basic institutions of family, religion, government, education, and economics.

A Content Model for the Study of Advanced Technology

In order to experience technological activity, students must use a process that is efficient and effective and that incorporates all the components necessary for technology to occur within the educational context. Therefore, the process model for the study of advanced technology must contain those components found in the technological method learning model (Figure 13). In essence, students must understand that human needs and wants lead to the identification of problems and opportunities that must be addressed using resources and technological knowledge in a systematic process to achieve evaluated solutions.

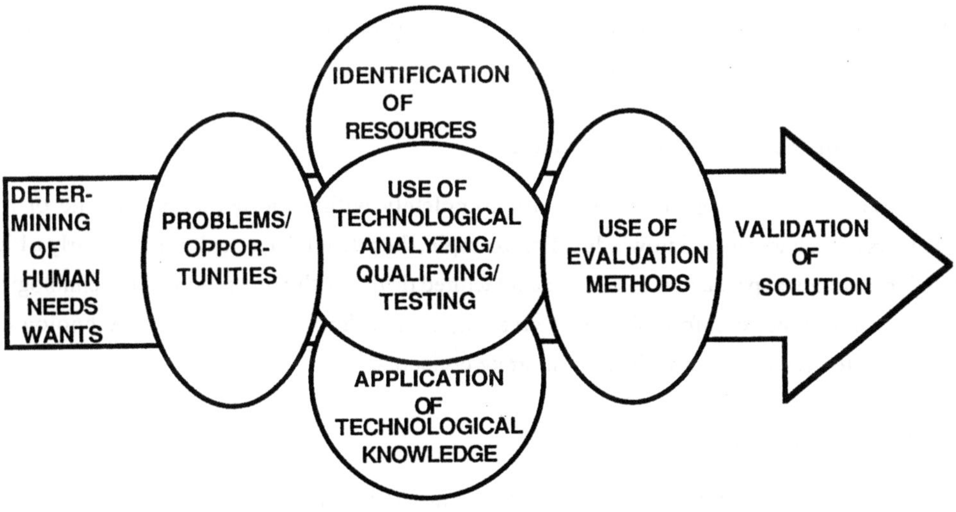

Figure 13. Technological learning model

Determination of Human Need and Wants

Human needs and wants must be determined from a productive citizenship level. According to Stephens [10], among the things implied by productive citizenship, in addition to occupational competence, are a well-developed value system compatible with the workings of society, a sense of responsibility to humankind, the ability to make decisions tempered with a sense of what is right as well as what is possible, a feeling of self confidence, the capacity for independent thought, and the ability to communicate effectively and efficiently.

Identification of Problems or Opportunities

An important element of attacking a problem or addressing an opportunity is to ensure that students "own" the problem, *i.e.*, they understand the problem and its relevance. Students must grasp the full meaning of the problem and begin to understand how to solve the problem. This can be accomplished through problem definition or goal setting.

Identification of Resources

Students must understand how all resources influence the selection of an optimum solution to any technological problem. Without viewing the limitations, restrictions, capabilities, and accessibility of personnel, tools/machines, data, materials, energy, capital, and time, proper management of technological ventures cannot occur. Constrains, or identified limitations, become a significant part of resource management. Such things as limited materials, finances, and time provide constraints. The laws of nature also provide constraints and preclude implementation of certain solutions.

Application of Technological Knowledge

Few outputs of technological ventures are completely within the taxonomy of one of the organizers of technological knowledge. For example, the artificial heart, although clearly a bio-related technological product, is in fact a manufactured device within the production technologies. Students should feel comfortable about crossing over technologies and using technological knowledge to create new models and integrate concepts.

Use of Technological Processes

The management of resources and application of knowledge must occur in a systematic fashion in order for technological ventures to be successful. Students must understand that all problems must be analyzed, realized, and tested. As these activities are accomplished, significant information is generated in the form of feedback to allow the student to modify any or all of the technological process actions.

Use of Evaluation Methods

Students must realize that a technological venture is not successful if the developed solution does not meet the goals, objectives, or criteria of the problem or opportunity. Evaluation is necessary to determine the level of satisfaction of the proposed solution.

Validation of Solutions

As the outputs of technological ventures occur, students must study the impact that those outputs will have within the historical, cultural, and environmental contexts. To teach technology void of these impacts would demonstrate a lack of sensitivity for the technology consumer. Yet there are many positive aspects of technology that must also be pointed out to students. People live longer because of improved medical technology and eat better because of household mechanization. A quality technology education program must include both positive and negative impacts as they relate to and interface with the technology systems of bio-related, communication, and physical technology. It must compare these issues to the personal values, norms, ethics, and beliefs of society over time and it must provide insight for decision making and technological problem solving now and in the future.

Conclusion

Advanced technological content must be process-oriented in order to accommodate the constantly shifting "state of the art" of technology. That process should contain and address all of the components of the technological method. To focus only on the organizers of technological knowledge or technological process is to take a simplistic view of the study of technology. Surely at the research level, specialization is desirable, even necessary. However, to study, teach, or learn technology, one must study, teach, or learn within the technological method because it is the process through which advanced technological content can be appropriately determined.

References

1. Boyer, E. *High School -- A Report on Secondary Education in America* New York: Harper & Row, 1983.
2. Frey, R. E. "Is There a Philosophy of Technology? Are There Different Approaches to the Philosophy of Technology? Is there a Framework for Critiquing a Philosophy of Technology? Is Philosophy of Technology Different from Philosophy of Science?" Paper presented at the meeting of the Mississippi Valley Industrial Teacher Education Conference, Chicago, IL, 1987.
3. Hannay, N.B. and McGinn, R.E. "The Anatomy of Modern Technology: Prolegomenon to an Improved Public Policy for the Social Management of Technology." *Daedalus,* (1), 25-53, 1980.

4. Hendricks, R. and Sterry, L. *Communication Technology,* 2nd ed. Menomonie, WI : T & E Publications, 1989.

5. Mitcham, C. "Philosophy of Technology." In P.T. Durbin (ed.), *A Guide to the Culture of Science, Technology, and Medicine* (pp. 282-363). New York: Free Press, 1984.

6. Savage, E. "Bio-related Technology: An Interview with Dr. Ernest Savage," *The Technology Teacher* (49) 8, 1990.

7. Savage, E. *Technology Systems Handbook.* Bowling Green, OH: The Model Technology Systems Project, 1990.

8. Savage, E. and Sterry, L. *A Conceptual Framework for Technology Education* (in press). Reston, Va: International Technology Education Association, 1990.

9. Skolimowski, H. "The Structure of Thinking in Technology," *Technology and Culture,* 7, 373-383, 1966.

10. Stephens, R. L. "The Role of Liberal Arts in Programs to Prepare Industrial Arts/Technology Teachers." Paper presented at the meeting of the Mississippi Valley Industrial Teacher Education Conference, St. Louis, MO, November 1986.

Bibliography

Bikjer, W. E., Hughes, T.P. and Pinch, I.J. (eds.). *The Social Construction of Technological Systems: New Directions in the Sociology and History of Technology.* Cambridge, MA: MIT Press, 1987.

Columbo, U. "The Technology Revolution and the Restructuring of the Global Economy. In Janet H. Maroyama and H. Guyford Stever (eds.), *Globalization of Technology* (pp. 23-31). Washington, DC: National Academy Press, 1988.

DeVore, P.W. *Technology: An Introduction.* Worcester, MA: Davis Publishing Company, 1980.

DeVore, P.W. "A Perspective of Technical Research." *36th Yearbook of the Council on Technology Teacher Education: Conducting Technical Research.* Mission Hills, CA: Glencoe Publishing Co., 1987.

Hales, J. and Snyder, L. *Jackson's Mill Industrial Arts Curriculum Theory.* Charleston: West Virginia Department of Education, 1981.

International Technology Education Association. *Technology Education: A Perspective on Implementation.* Reston, VA: 1985.

International Technology Education Association. *Technology: A National Imperative.* A Report by the Technology Education Advisory Council. Reston, VA: 1988.

Johnson, J.R. *Technology,* Report of the Project 2061 Phase I Technology Panel. Washington, DC: American Association for the Advancement of Science, 1989.

Kemp, W.M., and Schwaller, A. E. (eds.) *Instructional Strategies for Technology Education.* Mission Hills, CA: Glencoe/Council on Technology Teacher Education, 1988.

Market, L. R. *Contemporary Technology - Innovations, Issues, and Perspective.* South Holland, IL: Goodheart-Wilcox, 1989.

National Science Board Commission on Precollege Education in Mathematics, Science, and Technology. *Educating Americans for the 21st Century*. Washington, DC: National Science Board, 1983.

Pacey, A. *The Culture of Technology*. Cambridge, MA: MIT Press, 1983.

Science for All Americans - A Project 2061 Report on Literacy Goals in Science, Mathematics, and Technology. Washington, DC: American Association for the Advancement of Science, 1989.

Staudenmaier, J.M. *Technology's Storytellers: Reweaving the Human Fabric*. Cambridge, MA: MIT Press, 1985.

Swyt, D. "An Agenda for Progress in Technology Education," *The Technology Teacher*, 47(1), 3-8, 1987.

The Evolution of Current Practice in Technology Education

Robert McCormick

The Open University, UK

Abstract: This paper discusses how various reasons for teaching technology (economic, educational, citizenship, Marxisim) and traditions of teaching it (craft, design and art, science, STS, industry) account for currect practice in primary and secondary schools. The discussion looks back over the last hundred years at how the traditions have developed and interacted. A plea for a more sophisticated view of curriculum change is made.

Key words: technological literacy, curriculum change, curriculum history, technological awareness, craft, design, science

Introduction

Technology education has become recognized as a distinct area of education, and there is increasing international convergence of what constitutes technology education [I: 1, 16: I]. While there is convergence at the general level of policy documents, there remain differences of emphasis. These differences are illuminated by the way reasons for teaching technology and the way various traditions of teaching technology have combined in particular countries. I shall argue in this paper that:

a. we need to be aware of the various reasons for teaching technology, because they have provided support for the traditions and given rationale and agendas that come through in their practice;

b. we must learn from the various traditions because they encapsulate strongly held views and years of experience that will remain even after we have an established area of technology education.

I will also reiterate the plea made by McCulloch, Jenkins, and Layton [49:212] for a sensitive view of curriculum change and avoid the naivete of some who have tried to introduce technology into schools unaware of the forces exerted by the various interest groups that operate at both national and school levels. In making this case I will build on my earlier work [46] and that of Peter Medway [50].

My approach in this paper will be partly historical, but it will not be a detailed curriculum history, although I will draw upon such work [24,41,49]. Inevitably I face the danger of

NATO ASI Series, Vol. F 78
Integrating Advanced Technology into Technology Education
Edited by M. Hacker et al.

parochialism, but I have endeavored to be international in the literature drawn upon. My main endeavor is to draw out general points though they may not apply universally. Together the various traditions of teaching technology, and their associated reasons, encapsulate much of what is coming to be accepted as technology education. The problem is they have not yet come together sufficiently, and this paper attempts to help this process by clarifying the nature of, and the relationship among, the traditions.

Reasons for Teaching Technology

The reasons for teaching technology provide various interest groups, such as employers' organizations, science bodies (Royal Society or British Association for the Advancement of Science - BAAS), engineering interests, and teachers' associations, with the driving force to promote their view of technology education. The reasons for teaching technology that are dealt with here are those concerned with: improving the economy (the economic imperative); creating responsible citizens (citizenship); and educational benefits (intrinsic value). I will also consider briefly the political ideological reason that derives from Marxism.

The Economic Imperative

This reason is historically the most important, which can be seen in the arguments for manual training at the end of the 19th century. In England various commissions considering technical education contained industrialists who advocated manual training in an attempt to stem the decline of industry in the face of competition from the USA and Germany [41]. In 1979 Britain began the Great Debate prompted by the perceived failure of schools in the face of a declining economy. This led to the launch of the Technical and Vocational Education Initiative (TVEI) [49: Chapter 11]. This provided enormous funds and gave a boost to technology education, though of only a limited kind [56: 141].

In the USA at the end of the 19th century similar concerns were expressed about the economy, leading to recommendations for manual training to aid the development of American industry. During the 1980s concern over the state of industry has re-emerged, leading prestigious bodies to produce reports (such as Project 2061) and programmers to improve technology and science education [1, 2, 13, 31, 34, 53, 54]. In European countries with vocational or pre-vocational education similar economic justifications are used [29, 70].

Although improving the economy is a persuasive reason, support for technology education is complicated by a number of arguments about the relationship of education and economic improvement. I can do no more than list them here:

a. Vocational versus general education: Is vocational education better for economic development than general education? Probably not [57: 144-5]. Is the distinction between the two kinds of education valid?

b. Status of vocational education: What about the inequity of selection for vocational education and elitist views of academic education [63: 41-2]?

c. Science versus technical/technological education: How do the various science and engineering organizations and industry view the merits of each [49].

d. Do we change students' attitudes to industry or give them the skills and flexibility for future industrial technology?

These arguments complicate the case for technology education when it draws upon the economic imperative, and are a mixed blessing. In the UK, for example, employers have shown little interest in supporting vocational technical education in schools [48: 37, 63: 41], and have preferred science education rather than technology, albeit science with applications [49: 117]. They have, however, supported campaigns on improving attitudes to industry (*e.g.*, Industry Year in 1986), and have been more interested in school leavers' personal qualities than in their formal qualifications [26: 207].

The support in the USA for the approaches to education in the competing economies of the Federal Republic of Germany (as it was) and Japan [23: 71, 13: 2] ignores the particular stance adopted with respect to each of the above arguments in those two countries. Japan has little vocational education, very academic general education, and does no technology in schools [18,56]. Germany has well-developed vocational education, and technology is just being introduced into schools [58: 66].

Despite such unhelpful evidence, the economic imperative is seductive and has at times provided the motivation for funding by governments and industry.

Intrinsic Value

This reason for teaching technology is one which sees educational benefits by contributing something to an individual that other curriculum areas cannot. In the 19th century when industrialists were pushing for manual training to improve the economy, teachers and other educationalists were arguing for its moral, physical, motivational, and pedagogic benefits to students [41: 120-32]. This reflected their desire to raise their status [41: 367-9]. More recently the problem solving inherent in technology education has been seen as an educational justification, relying on Dewey's ideas on reflective thinking. The Internatioal Technology Education Association (ITEA) took this stance in the phrase "humans must act in order to be" [31: 8].

A second element in the problem-solving justification is that "real" problems require a multidimensional approach, which counters the fragmentation and artificiality of a subject-based curriculum. This, so the argument goes, is more in tune with the child's mind [19: 228-39]. This view is taken by those who support a progressive education ideology, and it has been instrumental in encouraging primary school teachers in the UK to take up technology education. A similar view is taken by some who support design education.

There is a further intrinsic justification: that it is necessary to learn technology problem-solving techniques, in the same way as we teach scientific investigation. This seems to be the implicit approach in the USA [3: 40, 20: 8, 72: 389, 53: 45]. Along with this is the concern for "doing" rather than "knowing," and for "knowing how" rather than "knowing that" [15: 29-31, 34: 4], reflecting tacit knowledge of the craft tradition. This concern has led to the idea of "capability," which in the UK has been supported by a group of influential people under the auspices of the Royal Society of Arts [61].

All this is a long way from the hand-eye justifications of a hundred years ago, but is nonetheless attributing the same intrinsic value to technology education as was done for handwork in schools.

Citizenship

This reason for teaching technology has two major overlapping strands. The first is a concern for the impact of technology and the role of the ordinary citizen in decisions about technology, in particular controlling it. The second strand concerns the need for people to survive in a technological world.

The concern for the impact of technology has its roots in the environmental movement associated with the "limits to growth" arguments of the 1970s [75: 535], and also with the "social responsibility for science" movement [68: 380]. These concerns were taken up in social studies and in science-technology-society (STS) courses in both the USA and UK, and early ones were anti-technology [75: 537].

The value-laden nature of technological activity and the need to reflect this in education are now accepted in most proposals for technology education [2, 17, 31], but not yet found in some areas of technology teaching [11]. Science, environmental, and social studies teachers are more likely to deal with value issues than craft or technical teachers.

The second strand -- preparing future citizens to survive in a technological society -- embraces the first strand, because it wants to prepare young people for all aspects of life that technology will affect. Understanding and using technology are important for this strand [53], and technological "awareness" or "literacy" are key terms in the UK and USA respectively [20,44]. Technological literacy may not be fully defined yet [9: (ii)], but its definition reflects two groups in the USA: STS supporters, represented by The National Association for Science, Technology and Society (NASTS), and technology educators, represented by the ITEA. They are not, however, entirely distinct, as Cheek acknowledged in his introduction to the international technology literacy conference that acts as the annual meeting of NASTS [9: (ii)].

Marxism

This has been a powerful force in the education systems of many countries. If it seems unfashionable at the time of the reunification of Germany, and not a little ironic in the context of a NATO-funded conference, then bear with me. Marx was concerned about the impact of large-scale industry in the 19th century, and his analysis of the position of the worker reveals some interesting points. He had remarkable foresight in seeing how craft models of work were being replaced, when 20 years later (in the 1880s) industrialists were advocating manual training in schools. To prevent workers from being enslaved by machines, he argued for mobility and flexibility to achieve a "totally developed individual" [43: 618], something advocated by present-day industrialists. To achieve this he advocated handwork and technological education [43: 619], which later become known as polytechnical education [66]. Thus we have an early recognition of technology education, and one which informs the practice of many of the Eastern European countries with which we will be increasingly involved [7, 52].

A second element from the Marxist tradition comes from ideas about the role of productive work in defining the nature of human beings, who create the world they inhabit and in so doing are changed by it. This parallels the ITEA view noted earlier (humans must act in order to be). Labor is therefore seen as important as part of education. In addition to this intellectual objective of productive labor there is the ideological one of wanting to overcome class differences and the distinction between mental and manual work. The argument about the status of technical versus academic education, noted under the economic imperative, has much in common with this latter distinction. This objective requires changes in attitudes to work, and again there are similarities with Western industrialists concerns, albeit from a different ideological standpoint.

A Mixture of Reasons

Those who have advanced the case of technology education have done so on the basis of one or more of the above reasons. The use of the reasons may vary over time and across the traditions that go to make up current practice. Thus science education has used economic, citizenship, and intrinsic arguments to justify its place, adopting a colonial stance towards technology. Similarly design education has called upon intrinsic and economic arguments, in the same way as manual training did at the beginning of the century. To some extent this may be opportunism, but also different factions within traditions, at different times, change the climate of rationales. The economic reason, although the most potent, causes the most difficulties as a basis for justification. For example, the use of the economic imperative to justify a technology program which emphasizes understanding technology in the world and associated social issues (as an STS course might) would not satisfy those who sought to improve manufacturing. An STS course is one possible response to the aims of the AAAS Project 2061: "Project 2061 has produced a blueprint which has a pure STS goal - Science for all Americans" [60: (i)].

The Traditions of Technology Education

Interest groups

I have already made the point about the various traditions forming interest groups, that, according to curriculum historians such as Goodson [24], determine the development of curriculum subjects. In this section I want to examine what each of the traditions contributes to technology education practice, relating it where appropriate to the conflicts and debates among the groups within each tradition and between the traditions themselves. These conflicts and debates at national level will shape technology education, and their adherents at school level will, along with the interaction of the practicalities of school life, reflect this shaping.

The traditions I have chosen to consider are: craft, design and art, science, science-technology-society, and those who are concerned for industry and productive labor. They are not an exhaustive selection, nor are they of equal concern in all countries.

Craft

This tradition is not a single one and differs across countries and time. Three strands are identifiable. One strand is (trade) craft stemming from the 19th and early 20th century manual training, which emphasizes exercises to develop skills with tools. Another form of craft developed from Swedish sloyd, that in some forms became aligned with art. The third strand is the development of manual training into industrial arts, with a concern for reflecting the developments in industry. There have been some interesting international exchanges of ideas, though not always with due consideration for the context. (See [5, 8, 22, 25,41] for accounts of the strands.)

Some advocates of Swedish sloyd based their case upon the child-centered approach of progressive education, and were concerned with free creativity, using Dewey's work as justification. Such advocates tended to ignore the social reconstructionist side of Dewey [65: 21] that Luetkemeyer sees as the foundation of industrial arts in the USA [42]. In contrast to the UK craft tradition, ossified in craft reminiscent of the 19th century [21: 5], industrial arts in the USA attempted to reflect industrial developments, although that did not continue into the middle of the 20th century [36: 4].

The difficulty for USA industrial arts, UK crafts, or for school vocational education now, as in the 19th century, is to reflect current industrial practice. This is the main context for most modern technological activity (*e.g.*, designing and making). The trade craft tradition has of course developed from wood to metal, from hand tools to machinery, but is unable to continue to develop as technology develops because the basic model of production is flawed and because there is a limit to what can be done in schools [50: 14-5]. The craft model of the single worker in control of the whole production process, that Swedish sloyd specifically sought to preserve at the advent of industrialization [5: 157], does not reflect the team work of industry. The failure of the use in schools of computer controlled machines, that ignores the context of a modern factory, is exacerbated as Flexible Manufacturing Systems (FMS) become more common in industry. A craft teacher used to the immediacy of the workshop sees production games and simulations as theory rather than practice. One response to this is to focus on the process through problem solving and design, an approach which picks up on the progressive education tradition of earlier times and is present in some English primary schools.

But the difficulties that the craft tradition may have should not blind us to the important contribution it makes to how to teach skill development, use materials, and manage individually based project work. The emphasis of this tradition on finished products does, however, create problems for the development of modelling skills.

Design and Art

The linking of these two is to some extent arbitrary, but to some extent a reflection of the British context within which design education developed. There are three main interrelated British interest groups (not all of them homogeneous): those from art and design, design education as part of general education, and craft, design, and technology (CDT). The interrelations are not altogether clear to me, as there has been less research on them than on other groups.

In the 19th century two groups existed within art and design [71]. One, the Society of Art Masters (SAM), was a hierarchical, subject-centered association of art school (male) principals dedicated to presenting drawing as an academic discipline, emphasizing classical draftsmanship and design allied to industrial arts. The other, the Art Teachers Guild (ATG), contrasted strongly with SAM. It was mainly made up of female classroom teachers and was egalitarian, individual-centered (child-centered) in its interests. Its outlook was to support those who considered the creativity, expression, invention, and imagination of childhood as important. This, Thistlewood argues, was a by-product of the English Arts and Crafts Movement, with its concern for the spiritual value of craftwork. Industrial art to them suppressed free creativity. In some senses the differences of SAM and ATG resembles the conflict of the manual trainers and the Swedish sloyd of about the same period, although I have no evidence to suggest that there was any connection. The successors to these two organizations combined in 1984 to form the National Society for Education in Art and Design (NSEAD), in a synthesis of their two approaches. Currently NSEAD sees itself as representing art and design interests to government [30: 123A], and in particular commented upon the National Curriculum for design and technology [55, 69].

Those seeing design education as part of general education focused initially around the Royal College of Art and Bruce Archer, who was responsible for an inquiry into design in schools [61]. He viewed design as a third area in education, distinct from but equal to the sciences and humanities [4: 49-56]. It concerns doing and making and includes such things as technology and fine, performing and useful arts; he represented technology as lying somewhere between science and design. Those who see technology as an independent area would put design and technology on a more equal footing [45: 7]. As an interest group they are represented by the Design Council and, along with other groups, are represented on the Design Education Forum. This Forum collectively impressed upon the working group drawing up the national curriculum on technology in schools the importance of seeing design, not technology, as the overarching curriculum activity [69: 11-12]. Such views have contributed to an understanding of design. In particular, they offer a sophisticated view of design based on what practitioners do, and given their art background, they not surprisingly want to emphasize imagination and intuition [55: 25]. They have also emphasized the importance of visualization, not just as a mode of aesthetic expression but as way of thinking, giving a sophisticated view of modelling [14: 104-21, 55: 25, 69: 12].

CDT developed under a number of influences, including that of design education [56], and the view of design which CDT has developed, epitomized by John Eggleston's work [21], has not surprisingly been criticized by the art and design groups. These criticisms include: a mechanical view of design, expressed through various linear and circular staged models that fail to teach pupils the process and do not reflect how designers think [33]; a view of craft that emphasizes an outdated woodwork and metalwork view of making, rather than one which develops new ideas and designs [55: 32]; and the problem of modelling I mentioned earlier.

Viewing these strands at school level, primary school teachers, for example, may find that the design education approach fits their progressive ideology, but may lack any appreciation of what design is like to the professional designer. They may align it with their views on self-expression in art, without the synthesis arrived at by NSEAD [71: 148].

Science

This tradition is only now trying to come to terms with providing "science for all," rather than for specialists [38: 15]. Having started with "applied" beginnings in the 19th century [37], science educators in England went through a post-war phase of emphasizing pure science, and were reluctant to support technology education [49: Chapter 6]. Now the Association for Science Education (ASE) recognizes the importance of technology education, as different from science education, and the role of science teachers in it [6, 64]. This development is mirrored internationally, though science education interests still seem to dominate [3, 38, 39, 73], as occurs in the mainstream science interests of the BAAS and the AAAS (American Association for the Advancement of Science).

The concern to make science relevant has increased the interest in studying science in meaningful contexts, technology being one of these. Problem solving therefore becomes important. Science brings to technology education: scientific investigation and experimentation methods including ideas of a "fair test" [6, 27]; a knowledge and conceptual understanding essential to a technologist; the development of new science and technology areas such as biotechnology, that CDT and home economics teachers, for example, are often ill-equipped to handle. Science educators have also built up an impressive empirical research literature from which we can learn and build upon to understand the unique problems in learning technology.

Science-Technology-Society

This tradition stems directly from the desire to educate for citizenship and to control science and technology, as I indicated earlier. It is one response by those in science education who want to teach science in context, to avoid the irrelevance of pure science, and by implication to replace "science for all" in schools by STS [76: 731]. Support among science teachers no doubt accounts for early STS courses' not distinguishing between science and technology, but now this situation is improving [40: 370-1]. The STS tradition has been responsible for developing value issues in relation to technology, and in particular has developed teaching material and methods for discussing controversial issues, role-play activities, and games [67: 272-3], methods more familiar to social studies teachers. This tradition also brings in a different kind of problem solving from that of design and make, one which relates to decision making about, for example, where to locate a power station.

Industry, Productive Work, and Education

In this subsection I will not adopt a historical perspective because industry and work is less a tradition than a common theme for a variety of groups. Marxist and Western industrial interests share some things in common, as I indicated earlier, and here I will try to illustrate the resulting practice. For both groups two interrelated aspects are important: an understanding of industry/ production and a change of attitude of students to it. Both take ideological stances, though of quite different kinds. Marxists are concerned to show the creative activity of production, along with an understanding and control over the production process (part of polytechnical education). Western industrialist argue for the importance of industry to wealth creation and the need to develop business and entrepreneurial skills and attitudes. Thus in England we have education for "economic and industrial understanding" as a cross-curricular theme [51].

Understanding Industry. For technology education the significance of understanding industry is twofold: to see industry as a manifestation of modern technological activity and to see industry as a part of culture, as a working environment. I have already noted the limitations of craft work as a model for modern technological activity. It may give experience of design and make, but it cannot illustrate properly the industrial processes. For example, at the heart of the current concern for quality in industry is process control, and the methods appropriate to it (*e.g.*, flowcharting and critical path analysis) are every bit as important as design, yet they are difficult to apply to craft-based activities, even individual student projects. Although the UK has a variety of support agencies, the tradition of using manufacturing, at least as a knowledge area, is better in the USA. It developed, as I noted earlier, from early 20th century analyses, and is one of the standard areas of technology originating from the Jackson's Mill analysis in 1981 [28]. Manufacturing was also singled out in the Project 2061 technology panel report as one of "the technologies" [34: 16-7].

The second significant feature of industry, as part of culture, is less well used in technology education. Work is a major part of life for most people and they ought to understand that part just as any other aspect of culture. Industries are complex organizations, with a variety of concerns from technical to environmental and economic. In a sense what we need is an STS of the inside of the factory rather than just looking at its external impact, as would be the case with environmental studies. The area of, for example, industrial relations has been more the concern of social studies or careers teachers rather than technology teachers.

Student Attitudes. Both Marxists and Western industrialists want students to have positive attitudes to industry, for different reasons. The experience in communist countries seems to be poor, although little measurement of attitudes to work has taken place. The only study I know of, in Zimbabwe, shows disappointing results for the government [10]. In the West the excellent studies of PATT, reported in the annual conference reports, *e.g.*, [12, 59], do not deal directly with attitudes to industry and have not to my knowledge yet looked at the effects on attitudes of activities that involve industry. Studies of introducing industrial applications into science teaching to make it more relevant and interesting are, however, not encouraging [35].

There are a variety of approaches to teaching about industry including games, simulations, role play, case studies, visits, and dealing with problems given by industry [11: 51-3, 32, 74], but they all have to face the difficulties of providing first-hand experience and realism. These are the central problems given the nature of modern industry, as I noted earlier with regard to FMS.

Learning from the Traditions

The above account of the traditions that have contributed to the evolution of technology education is obviously not exhaustive. Teachers of home economics parallel craft teachers in their development from cookery and sewing to food science and technology and textiles. Business studies teachers are now being brought in, especially when industrial and commercial aspects are concerned. Then there are the geography teachers for environmental issues! The potential list is long, and the selection of traditions I have made serves the purpose of showing the importance of traditions. No single one holds the answer to developing technology education, and it is doubtful if any one could take on the guise of all the others. Nevertheless there has to be a sharing of perspectives, always mindful that particular reasons for teaching technology will lead to different kinds of programs. Although I have been discussing these traditions at the level of national or international groupings, they have their correlates in schools. Most obviously they exist in particular subject and departmental groupings and in the interests, enthusiasms, and skills of individuals.

Curriculum Change

By way of a conclusion to this paper let me return to something I signalled in the Introduction: the need for a sensitive view of curriculum change. This view must recognize that it is not enough to draw up good proposals for technology education; the role of the interest groups that exist in support of or in opposition to technology education must also be taken into account. But that is only one level. As I noted at the end of the section on traditions, science, CDT/industrial arts/technical, art and design teachers will all share to some degree the characteristics of their respective traditions. Such traditions have been developing for a long time, and some have changed only slowly. As I have argued elsewhere [47], teachers will need time to develop their ideas about what technology is and their skills to teach it They will want to develop from their existing traditions, whatever mechanism for change is employed: the central dictate in the UK or the evangelism of the USA. Thus change will be a complex interplay of the national forces in professional and other interest groups, and their parallels in schools. Both levels must be attended to for the maximum effect of change.

Finally in the context of this international workshop we should remember that in the past there has been considerable interchange across the continents. Let us hope we can learn from history, and that we do not attempt to borrow parts of models or approaches regardless of the context within which they originated.

References

1. American Association for the Advancement of Science (AAAS). *The Continuing Crisis in Science Education: The AAAS Responds.* A Report to the Board of Directors. Washington, DC: AAAS, 1986.
2. American Association for The Advancement of Science (AAAS). *Science for all Americans: Summary,Project 2061.* Washington, DC: AAAS, 1989.
3. American Association for the Advancement of Science (AAAS). *Science for all Americans. A Project 2061 Report on Literacy Goals in Science, Mathematics, and Technology.* Washington, DC: AAAS, 1989.
4. Archer, B. "The Three Rs". In: A. Cross and R. McCormick, (eds.), *Technology in Schools,* pp. 49-56. Milton Keynes-Philadelphia: The Open University Press, 1986.
5. Arvidsson, H. "Swedish Crafts and Crafted Studies," *Design, Education, Craft and Technology*, 21,3, 157-161, 1989.

6. Association for Science Education (ASE). *Technology Education and Science in Schools.* Hatfield: ASE, 1988.
7. Blandow, D. *"The System of Polytechnical Education in the G.D.R." In: J. Raat, R. Coenen-van den* Bergh, F. de Klerk Wolters, and M. de Vries (eds.), *Basic Principles of School Technology*: Report PATT-3 Conference 1988, pp.116-35. Eindhoven: University of Technology, 1980.
8. Booth, B. "The Development of Technology Education in the United States," *Studies in Design, Education, Craft and Technology*, 21,2,84-9, 1989.
9. Cheek, D.W. "Introduction." In: D.W. Cheek and L.K. Walks (eds.), *Technological Literacy IV. Proceedings of the National Technological Literacy Conference*, pp. (i)-(ii). University Park, PA: National Association for Science, Technology, and Society, February 1989.
10. Chivore, B.R.S. "Education with Production in Zimbabwe: An Evaluation of Teachers' and Pupils' Attitudes and Perceptions," *International Journal of Lifelong Education*, 9,3, 201-218, 1990.
11. Clegg, A., Medway, P. and Yoemans, D. *Planning for Technology within the Curriculum Units 1-2, ET887/ 897, Teaching and Learning Technology in Schools.* Milton Keynes: The Open University Press, 1987.
12. Coenen-van den Bergh, R. (ed.): *Report - PATT-3 Conference 1987*, Vol. 2 Contributions. Eindhoven: University of Technology, 1987.
13. Committee on Education and Labor: *The Education Technology Act.* Hearing before the Subcommittee on Elementary, Secondary and Vocational Education, House of Representatives, 99th Cong., 2nd Series on H.R. 3102 (Feb. 19, 1986). Washington, DC: 1987.
14. Cross, A. "Towards an Understanding of the Intrinsic Values of Design Education." In: A. Cross and R. McCormick (eds.), *Technology in Schools*, pp. 104-21. Milton Keynes-Philadelphia: The Open University Press, 1986.
15. Cross, N., Naughton, J. and Walker, D. "Design Method and Scientific Method." In: A. Cross and R. McCormick (eds.), *Technology in Schools*, pp. 19-33. Milton Keynes-Philadelphia: The Open University Press, 1986.
16. Department of Education and Science and the Welsh Office (DES/WO). *Design and Technology for Ages 5 to 16: Proposals of the Secretary of State for Education and Science and Secretary of State for Wales.* London: DES\WO, June 1989.
17. Department of Education and Science and the Welsh Office (DES/WO). *Technology in the National Curriculum.* London: HMSO, 1990.
18. Dore, R.P. and Sako, M. *How the Japanese Learn to Work.* London: Routledge, 1989.
19. Down, B.K. "Problem-solving, CDT and Child-Centredness." In: A. Cross and R. McCormick (eds.), *Technology in Schools*, pp. 228-239. Milton Keynes-Philadelphia: The Open University Press, 1986.
20. Dyrenfurth, M.J. "The Technological Literacy Imperative Detailed." Prepared for the International Design and Technology Education Conference Series. Edinburgh: September 1990.
21. Eggleston, S.J. *Developments in Design Education.* London: Open Books, 1976.
22. Eyestone, J.E. "The Influence of Swedish Sloyd and its Interpreters in American Art Education." Unpublished Ph.D. Thesis. Columbia: University of Missouri, 1989.

23. Fowler, J.A. and Allen, G.G. *The Present Opportunity in Education.* A Position Paper on the Current State of Science and Technology Education, in the United States. College Park, MD: Triangle Coalition for Science and Technology Education, September 1988.
24. Goodson, I. (ed.) *International Perspectives in Curriculum History.* London and New York: Routledge, 1988.
25. Hacker, M. and Barden, R. "Systems Approach to Technology Education," *Man Society Technology*, 42, 6, 9-14, 1983.
26. Hargreaves, H. "Record Breakers?" In: P. Broadfoot (ed.), *Profiles and Records of Achievement: A Review of Issues and Practice.* London: Holt, Rinehart and Winston, 1986.
27. Harrison, M.K. "Science in Technology: Technology in Science." In: *Proceedings of the 3rd National Conference Design and Technology Education Research and Curriculum Development.* Loughborough: University of Technology, 1990.
28. Hayles, J.A. and Snyder, J.F. "Jackson's Mill Industrial Arts Curriculum Theory: A Base for Curriculum Derivation," *Man Society Technology*, 41, 5, 6-10, 1982.
29. Hochleitner, R.D., Artigas, J.T. and Cuerpo, M.G. *Education and Work in the Spanish Educational Reform.* A Study Prepared for the UNESCO Institute for Education in Collaboration with the International Educational Reporting Service. Paris: UNESCO, 1981.
30. Hughes, A. "The 1988 NSEAD Presidential Address: The Charge and Brief for Art Educators," *Journal of Art and Design Education*, 8,2, 123-33, 1989.
31. International Technology Education Association (ITEA). *Technology: A National Imperative.* Reston, VA: ITEA, 1988.
32. Jamieson, I., Miller, A., and Watts, A.G. *Mirrors of Work: Work Simulations in Schools.* Lewes: The Falmer Press, 1988.
33. Jeffery, J.R. "Design Methods in CDT," *Journal of Art and Design Education*, 9, 1, 57-70, 1990.
34. Johnson, J.R. "*Technology: Report of the Project 2061 Phase I Technology Panel.* Washington, DC: American Association for the Advancement of Science, 1989.
35. Labzonby, J.N. "Do Students Want to Learn About Industry?" In: D.J. Waddington, (ed.), *Education, Industry and Technology*, pp. 3940. Oxford: Pergamon Press, 1987.
36. Lauda, D.P. "Technology Education and Closing the Gap Between Theory and Practice," *The Technology Teacher*, 43, 6, 3-7, 1984.
37. Layton, D. *Science for the People.* London: Allen & Unwin, 1973.
38. Layton, D. (ed.). *Innovations in Science and Technology Education*, Vol. I. Paris: UNESCO, 1986.
39. Layton, D. (ed). *Innovations in Science and Technology Education*, Vol. II. Paris: UNESCO, 1988.
40. Layton, D. "Revaluing the Technology in STS," *International Journal of Science Education*, 10,4, 367-78, 1988.
41. Lord, W. "Manual Training in Public Elementary Schools, 1810-1910: A Study of Curriculum Change." Unpublished Ph.D. Thesis, Leeds: University of Leeds, 1988.
42. Luetkemeyer, J.F. "The Social Settlement Movement and Industrial Arts Education," *Journal of Epsilon Pi Tau*, 11,1-2,97-103, 1985.

43. Marx, K. *Capital,* Vol. 1. Pelican Marx Library Edition. Hamondsworth: Penguin, 1976.

44. McCormick, R. *Module 4 Introduction and Guide*. ET887/897 Teaching and Learning Technology in Schools. Milton Keynes: The Open University Press, 1987.

45. McCormick, R. *Teaching and Learning Design in Schools.* Module 1, Unit 8, ET887/897 Teaching and Learning Technology in Schools. Milton Keynes: The Open University Press, 1987.

46. McCormick, R. "Technological Education in the United Kingdom: An Inservice Provider's Perspective," Paper given to the World Assembly on Technology Education, International Technology Education Association. Norfolk, VA March, 1988.

47. McCormick, R. "Technology and the National Curriculum: the Creation of a Subject by Committee?" *The Curriculum Journal,* 1, 1, 39-51, 1990.

48. McCulloch, G. "City Technology Colleges: An Old Choice of School?" *British Journal of Education Studies,* 37,1, 30-43, 1989.

49. McCulloch, G., Jenkins, E., and Layton, D. *Technological Revolution? The Politics of School Science and Technology in England and Wales since 1945*. Lewes: The Falmer Press, 1985.

50. Medway, P. "Issues in the Theory and Practice of Technology Education." In: *Studies in Science Education,* 16,1-23, 1989.

51. National Curriculum Council (NCC): *Curriculum Guidance 4 - Education for Economic and Industrial Understanding.* York: NCC, 1990.

52. Navakova, H. "Basic Principles of School Technology in our Country." In: J. Raat, R. Coenen-van den Bergh, F. de Klerk Wolters, and M. de Vries (eds.), Report PATT-3 conference 1988, pp.108-15, Eindhoven: University of Technology, 1988.

53. National Science Foundation (NSF). *Educating Americans for the 21st Century: A Plan of Action for Improving Mathematics, Science and Technology Education for all American Elementary and Secondary Students so that their Achievement is the Best in the World by 1995. A Report to the American People.* Washington DC: NSF, 1983.

54. National Science Foundation (NSF). *Educating Americans for the 21st Century: A Plan of Action for Improving Mathematics, Science and Technology Education for all American Elementary and Secondary Students so that their Achievement is the Best in the World by 1995.* Source Materials. Washington DC: NSF, 1983.

55. National Society for Education in Art and Design (NSEAD). "Current Issues in Art and Design Education: the NSEAD Response to the Report of the National Curriculum Working Group for Design and Technology," *Journal of Art and Design Education,* 9, 1 23-37 (1990).

56. Penfold, J. *Craft, Design and Technology: Past, Present and Future*. Hanley: Trentham Books, 1988.

57. Psacharopoulos, G. (ed.) Special Issue: Vocational Education. *International Review of Education,* UNESCO Institute for Education: Iaamburg. 34,2, 141A8 (1988).

58. Raat, J.H., de Klerk, Wolters, F., and de Vries, M. (eds.) *Report PAT'T-3 Conference 1987.* Vol. 1 Proceedings. Eindhoven: Techniche Universiteit Eindhoven, 1987.

59. Raat, J., Coenen-van den Bergh, R., de Klerk Wolters, F., and de Vries, M. (eds.). *Basic Principles of School Technology*. Report PATT-3 conference 1988. Vol. 2. Eindhoven: University of Technology, 1988.

60. Roy, R. "Ten Years Young," *Bulletin of Science, Technology & Society*, 10,1, (i)-(ii) (1990).

61. Royal College of Art (RCA) *Design in General Education: The Report of an Inquiry conducted by the Royal College of Art for the Secretary of State for Education and Science*. London: Royal College of Art, 1979.

62. Royal Society for Arts (RSA). "Manifesto of Education for Capability," *RSA Newsletter*, spring (1984).

63. Sanderson, M. "Education and Educational Decline 1890-1980s," *Oxford Review of Educational Policy*, 4,1, 38-50 (1988).

64. Secondary Science Curriculum Review (SSCR) *Better Science: Making It Relevant to Young People*. Curriculum Guide 3. London: Heinemann/Association for Science Education, 1987.

65. Skilbeck, M. Ideologies and Values. Unit 3, *Culture, Ideology and Knowledge*. E203, Curriculum Design and Development. Milton Keynes: The Open University Press, 1976.

66. Smart, K.F. "The Polytechnical Principle." In E.J. King, (ed.), *Communist Education*. pp. 153-79. 1963.

67. Solomon, J. "The Dilemma of Science, Technology and Society Education. In: (Fensham, ed.) *Development and Dilemmas in Science Education*, pp. 266-81. London-New York Philadelphia: The Falmer Press, 1988.

68. Solomon, J. "Science Technology and Social Courses: Tools for Thinking About Social Issues," *International Journal of Science Education*, 10,4, 379-87 (1988).

69. Steers, J. "Design and Technology in the National Curriculum," *Journal of Art and Design Education*, 9, 1, 9-22 (1990).

70. Taylor, ME. *Education and Work in the Federal Republic of Germany*. London: Anglo-German Foundation for the Study of Industrial Society, 1981.

71. Thistlewood, D. "The Formation of the NSEAD: A Dialectical Advance for British Art and Design Education," *Journal of Art and Design Education*, 8, 2, 135-152 (1989).

72. Todd, R.D. "A Conceptual Basis for a Technology Education Curriculum." In: R. Conen-van den Bergh (ed.), *Report PATT conference 1987*. Vol. 2 Contributions. Eindhoven: Technische Universiteit Eindhoven, 1987.

73. UNESCO. *Science and Technology Education and National Development*. Paris: UNESCO 1983.

74. Waddington, D.J. (ed.). *Education, Industry and Technology*. Science and Technology and Future Human Needs - Vol. 3. Oxford: Pergamon Press, 1987.

75. Walks, L. "Afterword: New Challenges for Science, Technology and Society." In: D.W. Cheek and L.J. Walks (eds.) *Technological Literacy IV. Proceedings of the National Technological Literacy Conference*, pp. 535A5. University Park, PA.: National Association for Science, Technology, and Society, February 1989.

76. Za'rour, G.I. "Forces Hindering the Introduction of Science, Technology and Society Education in Schools." In: K. Riquarts (ed.), *Technology Education: Science-Technology-Society. Science and Technology Education and the Quality of Life, Volume* 2. Papers submitted to the International IOSTE Symposium on World Trends in Science and Technology Education, pp. 731A1. Kiel, West Germany: Kiel University, Institut fuer die Paedaggik der, 1987.

Contemporary Technology: A Philosophic View

Ronald D. Todd

Drexel University, USA

Abstract: Scholarly treatment of technology has been neglected, resulting in inadequately developed language, conceptual models, and frameworks of technology. Lack of a common language and conceptual portrayals of technology has made the international sharing of efforts in technology education difficult. The inadequate conceptualization of technology has also made it difficult to identify relationships between technology education and other school subjects, notably the sciences, mathematics, and the social sciences. Finally, the inadequacy of such constructs makes it difficult to identify the essential knowledge and skills of technology and how these would apply to contemporary and advance technology programs.

Key words: definition of technology, philosophy of technology, frameworks, paradigm, science, technological method, design, problem solving

Introduction

Technology as a form of human endeavor is arguably as old as civilization. However, technology has become an area of academic interest only during this century, and technology as a school subject is but a few decades old. Technology as a subject is so new that its content and methodology remain largely undeveloped. Consequently, those who would teach technology represent a field marked with uncertainty and confusion.

Within this context of uncertainty, I will consider three problems affecting technology educators in their efforts to develop contemporary technology education programs. First, it appears that relatively few scholars and philosophers have seen technology as a subject worthy of much attention or study. Second, lacking conceptual clarity and solid philosophical underpinnings, technology education has evolved guided not by philosophy but by operational paradigms. Third, the continuing confusion about the essence and essentials of technology makes it difficult to develop programs appropriate for the present, much less the future.

NATO ASI Series, Vol. F 78
Integrating Advanced Technology into Technology Education
Edited by M. Hacker et al.

Philosophy and Technology

Historically technology has generally been given little attention by academics, but nowhere is that neglect more surprising than in philosophy. Feibleman observed that both science and technology have been neglected by philosophy in quite different ways but that technology had been ignored altogether.

> Throughout history, technology has played a crucial part in human culture but there is no record that philosophers took any account of it. They give a lot of attention these days to the existence of technology but very little to the products of technology, but in no major writings of the classical philosophers has the existence of technology even been mentioned [5].

There now is a growing body of philosophical works regarding technology, but confusion persists in the search for clarity and meaning of the concepts and components of technology. Part of that confusion arises because of what Rapp describes as a lack of differentiation between discourses that are more "technological philosophy" than those that are "philosophy of technology." "Technological philosophy ... explicitly refers to 'the people concerned in producing and making use of technical objects'... and ... is part of a broadly conceived social and cultural philosophy" [11:XII]. The philosophy of technology approach turns attention away from the human element and attempts a more objective view appropriate to a general methodology and theory of knowledge.

Both types of philosophical pursuits are important but, for the curriculum design and development responsibilities faced by technology education professionals, efforts related to the philosophy of technology are essential. The theoretical products from the philosophers of technology can help provide and strengthen the conceptual frameworks that guide our ongoing curricular work. Lack of such products will hamper these curriculum development efforts significantly. Alternatively, professionals from technology education must turn their attention away from developing conceptual frameworks and curricular products and attend to developing the needed theoretical constructs -- a task for which most in the profession are ill prepared.

The relative paucity of philosophical treatment of technology exacerbates an already difficult situation. There is no readily identified group of professionals called technologists to be drawn upon by those in technology education. Unlike science education, which can turn to an almost unlimited number of professionals called scientists, technology education has far fewer and less apparent options. Individuals who fill the role of technologists tend not to think about technology in the same way that scientists think about science. Professionals such as engineers and architects, who study and use technology and consequently might qualify as technologists, tend to be offended by the designation. Consequently, the relative lack of attention given by philosophers to technology is very serious for technology education as an emerging profession.

This paper is based on the premise that technology education should be about the teaching and learning of technology. If those engaged in technology education must turn their attention to clarifying and describing what technology is, time and attention will be taken away from clarifying and describing what technology education is. A cloudy vision of technology has resulted in a cloudy vision of technology education. Consequently, instead of technology we have substituted production, communication, transportation, biotechnology, systems analysis, design and problem solving, applied science, history of technology, high technology, and a plethora of specific technical skill courses. Our dilemma is a serious one. As we face the continuing escalation of knowledge generally and technological knowledge specifically, we must be confident that our students are gaining the essential insights, capabilities, and beliefs related to technology. Few of us have that confidence. All of us are concerned that our graduates will have the knowledge, skills, and attitudes needed to participate in and contribute to a world shaped technology.

Meanings of Technology

The study of technology is very young and therefore is fraught with a diversity and ambiguity of ideas, directions, and meanings. Emerging from many of the definitions of technology is the idea of putting knowledge to use in order to create something. This idea, put forth by Skolimowski [13], has been echoed by Harrison [17], Todd [14], and DeVore [3]. Mesthene [8] equated new technology with change that often results in creation of new options.

Problem solving as a key aspect of technology is addressed in the English document *Technology in Schools*. Here it is suggested that technology is concerned with solving problems where there are not right or wrong answers but good rather than bad solutions.

> Technology is concerned with the identification of the needs of man and his endeavor to satisfy those needs by the application of science and use of materials resources and energy. It is concerned with solving problems where there are not right or wrong answers only good or bad solutions to a problem [2:180].

Definitions of Technology

One could question whether it is possible or even desirable to identify a single definition of technology. Multiple meanings need not be seen as a failure of the profession to sort out its thinking. Rather, multiple definitions could suggest the richness of meaning ascribed to the term. The categories identified below are provided as a way of looking at some of the multiple meanings of technology. This set of categories represents a perspective on technology as knowledge put to use for the purposeful creating of something practical. They include the following:

product: a tangible result of purposeful action; an artifact or article, *e.g.*, a computer, abacus, laser beam, or stone ax.

process: the practice by which the artifacts and products are generated; the operations involved in producing or making things.

procedures: the methods used to guide the processes, *i.e.*, the application of knowledge and skills to solve practical problems.

knowledge: the capacity, power, or ability of application, *i.e.*, the knowledge and skill derived from and applied in the use of products, processes, and procedures.

For the purposes of this paper and ensuing discussion the following definition of technology is stipulated. *Technology is a body of knowledge and the application of resources using a systematic approach to produce outcomes in response to human needs and wants* [12:3].

Philosophic Bases for Technology Education

There has been a great deal of "philosophizing" about technology education that spans nearly 50 years of professional discourse. As indicated earlier, much of that discourse has been more technological philosophy than philosophy of technology. The shaping of technology education as a profession has also been guided more by operational paradigms than by articulated and accepted philosophies. The seeds of one of the dominate paradigms in the United States were found in the document, *A Curriculum to Reflect Technology,* developed at Ohio State University in 1946 and 1947 under the leadership of William E. Warner [10]. The ideas and philosophy underlying that document, depicted in the diagram below, were to represent major influences on the thinking of those in the United States who would attempt to give technology education both meaning and utility. The work of Delmar W. Olson, especially his book*ndustrial Art andT echnology*, can be seen as the next generation of ideas concerning technology as a subject to be learned [9]. Olson was the first to grapple with the difficult problem of identifying the new content structure of industrial arts if it were, indeed, to reflect technology. Most of the work that followed treated technology as a part of industrial arts, and culminated in 1980 in the form of the *Jackson's Mills Industrial Arts Curriculum Theory* [6].

A Curriculum to Reflect Technology

The Communications Division

Composition & Duplication	Tranmission & Reception	Interpretation
Graphic Arts, Sound Recording	Mechanical-Electrical	Visual Sounds & Codes
Drawing, Sketching	Telegraph	Historical
Drafting, Blueprinting	Telephon	Signal Flags

Letterpress	Radio (CW<MOD)	Lights
Photography	Teletype	Sound Devices
Intagliography	Facsimile	
Planography	Television	
Duplication	Multi-Channel Methods	
Sound Recording	Radar	

The Construction Division

Homes
Highways (including Bridges and Tunnels)
Factories and Public Buildings
Airports
Waterways
Single fabrication, housing, public works, industrial, national defense...

The Power Division

Sources:	Natural, Electrical, Thermal
Generation:	Solar, Hydro, Biological, Combustion, Nuclear Fission, Electrical
Transmission:	Hydraulic, Pneumatic, Mechanical, Electrical
Utilization:	Manufacture, Construction, Transportation, Communications

The Transportation Division

Land:	Highways, Railroads
Air:	Heavier than air, Lighter than air, Navigation, Meteorology, Airport Aerodynamics, Space
Sea:	History Carrier types, Ship construction, Power plants, Propulsion units, Small boat building, Model making, Terminals, Routes, Organizations, Documents

The Manufacturing Division

Major Areas:	Food, Textile, Rubber, Chemical, Cellulose Fiber, Leather, Metal Ceramic, Miscellaneous
Areas of Study:	History, Materials, Fabrication, Consumption, Applications

Due in part to the direction set by Warner and Olson, the overriding concern in the United States was the content to be taught in American schools. Only a modest amount of attention was turned to process generally and the process of technology specifically. In the United Kingdom attention was turned more to that process of technology with far less attention to what technology was and what content it represented. One paradigm that appears to have been very influential in the development of technology as a school subject in the United Kingdom is presented in Figure 1 [7]. To illustrate the durability of paradigms, this model has recently been revised to include design and offered as a Model for Design and Technology in the new National Curriculum.

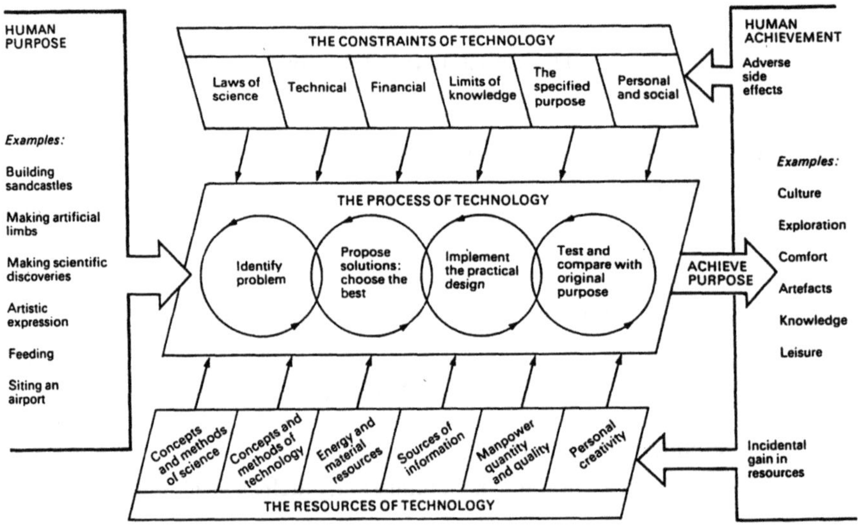

Figure 1. Technology -- A disciplined process using scientific, material, and human resources to achieve human purposes

Internationally, other paradigms have emerged to guide the development of technology as a school subject. In selected countries, such as The Netherlands and Germany, the efforts have been significantly influenced by scientific perspectives. In others, the dominant perspective or paradigm for technology education has been drawn from vocational education. The differences in the paradigms used by different countries, and consequently their specific operational forms of technology education, may be influenced by the level of technological development of those countries [15]. One recurring model that may become the new paradigm, or will certainly influence the current paradigm for technology education, is the general systems model shown in Figure 2.

Not unexpectedly, the meanings, forms, and results of technology education internationally now range over considerable territory. The emerging potential of technology as an arena of professional endeavor has contributed to that diversity of thought and richness of ideas. More central to our concern has been the lack of balance in developing the philosophic underpinnings

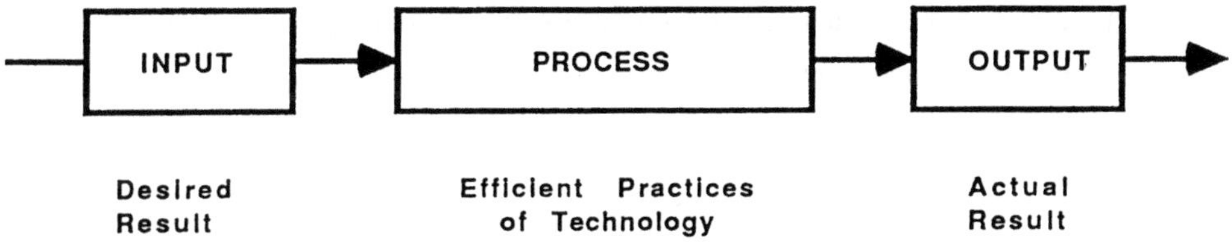

Figure 2. General systems model

of technology education. For discussion sake, let us consider that significant aspects of those underpinnings are drawn from the theories and methods of epistimological, praxiological, and axiological endeavor. We tend to adopt one of these as a perspective and slight the other two. What we need is a balance across the three. From epistimological pursuits can come insights as to what knowledge is enduring. Praxiological thinking can help identify skills that are enabling, while axiological efforts may suggest values and attitudes that are liberating. All three are important, even essential, if we are to help our students keep up with the advancing technology.

Technology Education as an Emerging School Subject

As indicated above, technology education is a subject area in search of itself. That search is made more difficult because, as observed by de Vries "technology is a school subject without tradition" [4:27]. The lack of a long-standing and rich tradition underscores the importance of increasing collaborative efforts across the profession. More specifically, we must continue to develop networks, mechanisms, and resources that support international sharing. The dramatic crumbling of physical and political walls that have separated family members of the technology profession sets a new and exciting stage for collaborations -- some of which are already underway.

As a case in point, the efforts in the United States by the Technology Education Task Force have been enriched through collaboration with colleagues from countries including the United Kingdom, the Netherlands, Australia, and Germany. That collaboration has helped to draw together different perspectives on content, process, and attitudes of technology and may well result in a new paradigm for technology education in the United States that is currently described as the technological method.

The Technological Method

The technological method, as presented in Figure 3, represents a model of "doing" technology. By attending to human needs and wants (1) problems and opportunities (2) can be addressed by applying resources (3) and technological knowledge (4) through technological processes (5).

The result of this effort can be evaluated (6) to assess the solutions and impacts (7) resulting from these general technological activities.

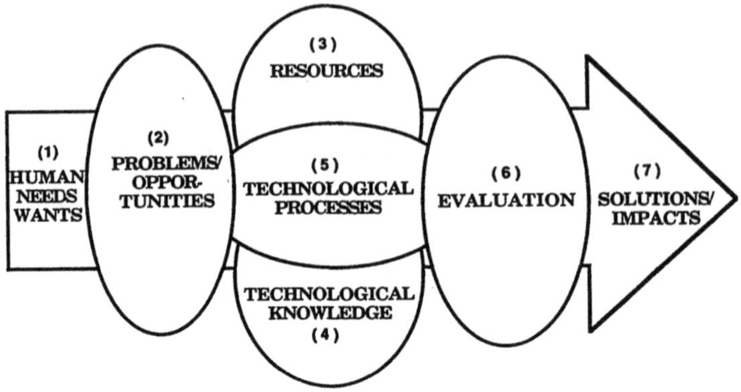

Figure 3. The technological method

(1) *Human Wants/Needs*

Technological endeavors should be in response to human wants and needs. Consequently, students should learn how to identify needs and wants through means such as observation, interviews, and research.

(2) *Problems/Opportunities*

"Problems are instances of uncertainty and ambiguity that actually exist and can be identified and described. Problems call for reactive behaviors on the part of the participants. Opportunities are possibilities that might arise or might be created in the future. Opportunities provide a much wider range of responses than problems and call for proactive behaviors from the participants.

(3) *Resources*

Resources represent the basic requirements for implementing or completing the technological processes. The resources include materials, energy, information, humans, tools/machines, capital, and time. The first three resources --materials, energy, information -- are the basic stuff from which all technological artifacts are made. This is accomplished over time by humans using tools and machines. Some form of capital is needed in order to acquire or assemble the resources that will be applied to the problem.

(4) *Technological Knowledge*

Technological knowledge is that body of facts, concepts, procedures, principles, and generalizations that can be applied to practical use. Such knowledge is found in each part of the technological method and within the categories of bio-related, communication, production, and transportation technology. It should be noted that these categories, although useful, are seen by some in the profession as imposing serious limits on the categories of what is considered technological knowledge.

(5) *Technological Processes*

Technological processes involve participants in (a) analyzing, (b) realizing, and (c) testing. (a) Analyzing is an important step in determining needs and wants and translating them into problems or opportunities. Analysis leads to the process of gathering information and generating alternative solutions in response to a problem or opportunity. Information gathering helps provide insight into the nature of the circumstances, development, and characteristics of the problem. The information provides a foundation from which alternatives can be development. Alternative solutions are generated through creative and inventive processes. (b) Realizing is the process of translating thought into action resulting in some form of physical reality. Through realizing the most appropriate of the alternative solutions is selected and developed as a prototype. The selection of the most appropriate solution is based upon the requirements and specifications of the problem. The development of a prototype provides a physical reality representing the proposed solution. (c) Testing is the process for determining if the outcome of the technological process works and how well it works. The testing is based upon the criteria and specifications set for the product. Feedback from testing may indicate that the products should be modified and improved or that aspects of the analysis or realization effort need to be modified or replaced.

(6) *Evaluation*

Evaluation builds upon but goes beyond the testing aspects of the technological processes. Evaluation in this larger sense requires that the process and results of the technological method be assessed for their acceptability. The results or products would be assessed against the criteria, specifications, and requirements set in the problem statement and the related design brief. The different steps within the actual process would also merit evaluation to determine what aspects of the overall process might be modified and improved. Insights gained through evaluation may require the products developed and/or the procedures to be changed.

(7) *Solutions and Impacts*

The evaluating of the products and processes sets the stage for helping students consider the solutions and impacts of their activities. In a sense all technological activities and decisions can result in solutions, whether adequate or inadequate. Since all technological solutions have impacts, one means of determining the adequacy of a solution is to determine what kind of impact the solution will actually have [12].

It is very important that we help students see that all technological decisions have consequences. Such consequences may be intended or unintended, desired or undesired, as indicated in Figure 4. Additionally, some consequences may become apparent on an immediate or short-term basis while others may be delayed and not emerge until some time in the future.

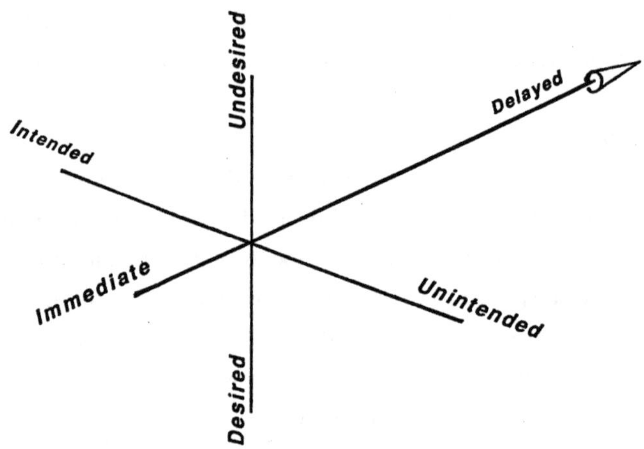

Figure 4. Impacts and consequences of technology

Turning attention to the impacts and consequences of technology provides an important connection with other school subjects. Many of our colleagues in education will feel uncomfortable and unprepared to practice technology, but they will often have some background and certainly the interest to discuss what technology is doing to them and to their world. Each connection represents an opportunity to enrich the learning of teachers and students in their study of technology and other subjects as well.

Technology Education and Other Subjects

Resided in the meanings attributed to technology are the recurring components of knowledge and application. It is common to hear and use terms such as applied science, applied mathematics, applied writing, applied psychology, and the like. Some individuals are, indeed, willing to accept the concept of applied science as equivalent to technology. A case can be made that application is important to technology and consequently important to technology education. Accepting that application is directly and integrally linked with technology poses important questions. Does the application of knowledge move only toward technology or is there a flow of knowledge from technology to other subjects? Is there an applied technology related to science in math or psychology? If so what is the nature of that relationship? If the knowledge of technology is not applied in other subjects, what other relationships, if any, are there? The compendium of questions continues to grow far beyond what is possible within this paper. Consequently, I would like to share with you some thoughts on the working relationships between science and technology [16].

Technology and Science: An Example

For discussion's sake let us consider that technology and science represent uniquely different yet mutually supportive bodies of knowledge and methods of discovery. Science is about knowing. Its major goal is to understand nature and its operation through the accumulation of empirical knowledge rather than other forms of knowledge. Although viewed by some as the source of "truth" and a body of "facts," science is more appropriately viewed as a source of models and theories that attempt to explain not "what is," but "what we know" and as a method for expanding what is currently "known." Historically we have seen usable concepts and ideas that were later found to be misleading or wrong. From this perspective, science is a collection of concepts and theories that help us understand and deal with nature and a method of inquiry that can be used to expand that body of knowledge.

While science is concerned with knowing, technology is essentially concerned with doing or more accurately a specific type of doing. Technology starts with an idea or need (sometimes drawn from science, but often not) which is converted through the procedures and processes of problem solving into physical reality. Technology does not exist until that transformation takes place. Science, on the other hand, is essentially a body of ideas and does not generally require a transformation from thought into action. The exception comes when theories must be put into practical application in order to test their validity.

Within science, there is modest interest in putting ideas into application. Technology represents the domain of application. Individual scientists may have applied interests, and in such cases, they literally transform themselves from scientists to technologists. Scientists use technology regularly, and without technological support, science would be basically deaf and blind. Technology provides the instrumentation both as hardware and software through which scientists can observe phenomena beyond the range of their unaided senses. Technology also provides the hardware that enhances the scientist's capacity for memory and creativity; the computer is perhaps the most evident example.

Technology and science in the real world may be quite different and the efforts of science and technology often remain separate yet supportive of each other. Technology often provides the tools and devices that extend the senses so that scientists can test their ideas. This technology is such an integral part of the work done by scientists that, for all practical purposes, the technology becomes invisible. Similarly, science provides a theoretical backdrop that makes the training and work of technologists more efficient and productive. Science, through knowing in this sense, helps technology to continue its business of doing. Similarly, part of doing for technology is to support science's major purpose of knowing.

Identifying the nature of the links that exist between science and technology is important, particularly if we are to enhance the teaching of the two subjects. To that end the two fields of endeavor and knowledge are represented in Figure 5. The overlapping circles indicate that there is a common arena of knowledge and activity where science and technology are integrated. This area represents a well-developed body of knowledge that shows the connections between the two subjects. The larger part of each circle represents the activities and knowledge that can be considered unique to science or technology. The separation of these two areas indicates a lack of specific links or that such links have not been identified or developed.

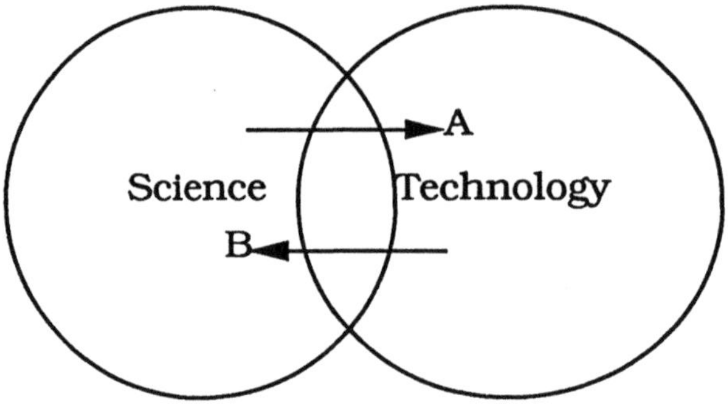

Figure 5. The relationship of science and technology

Within the diagram the arrows indicate major links between science and technology that could influence the teaching of the two subjects. The first arrow represents what science provides to strengthen and extend technology. This includes the scientific constructs of laws, principles, and concepts that provide insight into the scientific underpinnings of technological activities and phenomena. For example, science provides Newton's laws that in turn provide insight into technological instances entailing the movement of mass. Such instances include the operation of machines, transporting by trucks and trains, and the design of new land, sea, air, and space vehicles. Additionally science provides knowledge that is helpful in developing new materials that can be used for new products, structures, and inventions. Science also provides principles and insights that help develop new processes for the more efficient and useful conversion of old and new materials.

The second arrow represents the instruments and devices that extend the senses and capabilities of scientists as they engage in exploring natural phenomena. In essence technology provides the "tools" of science so essential to the ongoing effort of expanding the boundaries of "knowing." Technology also provides new inventions that lack scientific explanation of their operation. These instances cry out for an explanation and will often trigger intellectual efforts that lead to new insights and knowledge. For example, the steam engine was built and used for years before the underpinnings of scientific theory and concepts of thermodynamics were developed. More recently technological advances in lasers, superconductivity, and computers have jumped ahead of scientific understanding. Such symbiotic progress sets the stage for further research that helps extend the knowledge and practice of science and technology alike.

Matter and Energy and Information

As an example of the type of work that lies before us, I would like to build upon some ideas put forward by James Feibleman in the early 1980s. Feibleman suggests that philosophy has been changed drastically by science and that among these radical changes, none is more important than the idea that matter is the primary reality. Materialism within philosophical thought has been around at least since the Greek atomists who considered the "atoms" as round bits of impenetrable and unchangeable stuff. The atomists viewed these atoms as falling, with heavier atoms falling faster than light ones, and when hitting each other in an off-centered manner causing sideways motion. [5:18].

Both of these concepts have limited explanatory value that became transparent in the 20th century when enormous strides were made in discovering the structure and properties of matter. These discoveries were made possible because of the rapid advances in technology, like instruments to probe the atom and discover its constituents. Speaking generally, technology is responsible for the devices which use energy to move matter or alter it in various ways. As a consequence matter is no longer regarded as simple, inert stuff which resists analysis but has come to be recognized as a highly dynamic agent capable of sustaining the most complex of activities. It can no longer be considered separate from energy [5:18]

Matter and energy are best thought of as two forms of substance, with matter seen as static substance and energy as dynamic substance. As described in Einstein's most famous theory, $E=mc^2$, the two are interchangeable. This departed significantly from Newton's thinking, which saw energy as external to matter. Feibleman describes energy as a property of matter and that matter can be transformed into energy which in turn exerts force. In this sense matter is potential energy.

Let us now turn our attention to a missing part of the puzzle. Information has emerged as a key aspect of technology education. Information was treated as a key idea by Wiener in his work on cybernetics. Wiener made a very important philosophical statement related to energy and matter but focusing specifically on information. Wiener proposed that information was neither matter nor energy but something uniquely itself [18]. That statement represents the seeds of a paradigm that shaped and directed the emergence of a new fields of scientific inquiry and technological endeavor -- cybernetics and information theory.

Those of us in technology education, however, have given only modest attention to the concept of information and what Wiener's statement might actually mean. In simplified form information can be treated as patterns of meaning. But what forms can this patterns take? Let us consider several examples. Thousands of years ago astronomers were able to identify patterns of movement of the planets and stars that in a sense provided meaning. It would be easy to argue that much of the meaning was proven inaccurate or even wrong but that is a different argument for a different time. More recently the insight gained concerning the structure of DNA has provided and is still providing patterns of meaning of immense importance to us all. The DNA molecule not only provides information that supports inquiry into growth, illness, and genetics but also provides a means of establishing links across time and space with prehistoric human ancestors and forebears. Researchers are currently attempting to identify and decode the information that may be locked into rock formations as a means of determining when and where earthquakes might occur. Finally, the work within information science has yielded insights into the nature of messages that are transmitted in the form of signals of energy such as sound, light, or electricity.

Working with teachers, my colleagues and I have found it useful to define information as "patterns of energy and material that infer meaning." This stipulated definition does justice to the idea that matter and energy are the two essential substances within the universe and also attends to information as an important related, if not equal, concept. It was necessary, however, to expand the definition to consider the uses and forms that the information would take. The expanded definition would be "patterns of energy and material, as signals and symbols, that infer meaning for purposes of communication and/or control."

Information is seen as one of the three pillars of technology in The Netherlands as observed by van der Velde [17:426]. Similarly, information is given status equal to materials and energy by Bandlow in his conceptualization of technology as a school subject [1]. The British identified material (matter) and energy as key concepts and included information as a part of control in their early curricular thinking. Lately, with the emergence of the national curriculum, information has been elevated to a higher level of importance while energy as a concept appears to be given less importance than before. In North America, particularly the United States and Canada, there has been little agreement on the basic elements of technology. Only recently has there been a movement toward some professional consensus as to what some of the essential components of technology might be.

That brings us to the important question: What is the nature of advanced technology? Is advanced technology more difficult than standard forms of technology? Is it more theoretical, more mathematical, or is it primarily the new and novel? Lacking any sense of the history of technology, the uninitiated may well equate new developments with advanced technology. This can create some very real problems. For example, robotics -- an accepted instance of advanced technology -- is not something that has happened within the last two or three years. The roots of robotics can be traced back several hundred years to past instances of automata or as far as two thousand years to the early control devices invented by the Greeks. What is it that makes robotics simultaneously old and advanced?

Is advanced technology represented by being more theoretical or mathematical? Again, from an historical perspective, who is to say that the microelectronics of today are any more difficult or more advanced than the steam engine in its day. At the risk of using a terrible pun, the steam engine can be seen as the driving force for thermodynamics as a whole new field of scientific inquiry whose latent roots can be traced back to the discovery and application of fire. Is advanced technology the complex technology of its time? These and other questions must be faced if itechnology education is to be established as an emerging and important school subject.

Closing Observations

The important point that I have been trying to make throughout this paper is that we have a great deal of work ahead of us. There are literally hundreds of points of confusing language, concepts, and relationships, similar to the information example above, that need clarification. This work is essential if we are to talk a common language of technology and if we are to move beyond the threat of politically inspired curricula. Perhaps this idea is based more in fantasy than reality since curriculum development is always a political endeavor. Without well-developed ideas, clear and powerful concepts and a language of common meaning regarding technology, as a profession, we will be lost within our own tower of Babel. With such a foundation of common meaning, a new level of international cooperation and sharing will be possible, and international collaboration is a prerequisite if we are to achieve the daunting task of identifying the essential knowledge of technology that must be conveyed to our students.

Our knowledge relative to technology generally and technology education specifically should be viewed as transitory rather than enduring. Unfortunately there is no well-established method of "scrapping" knowledge when it becomes unusable. One alternative is to direct our attention to the identifying the knowledge of technology that will be enduring and transferable. That task will certainly be demanding, and we should encourage others in our field to engage in this effort. Fortunately the number of academics, including philosophers, interested in technology appears to be increasing. More importantly, we will want to listen to what these colleagues have to say. It certainly will be to our own best interest.

References

1. Blandow, Deitrich "The Critical Starting Point for Effective Technological Instruction." Mimeograph, Erfurt, Germany: Pedagogical College 1990.
2. Cross, Anita and McCormick, Robert. *Technology in Schools.* Milton Keynes, England: Open University Press, 1986.
3. DeVore, P. *Technology: An Introduction.* Worcester, MA: Davis Publications, 1980.
4. de Vries, Marc "The School Subject 'Technology': A Subject Without Tradition" In: PATT (Pupils' Attitude Toward Technology) Conference Report, p. 27. Eindhoven: Technische Universiteit, 1987.
5. Feibleman, James K. *Technology and Reality.* The Hague: Martinus Nijhoff Pubs., 1982.
6. Hales, J. and Snyder, J. (eds.) *Jackson's Mills Industrial Arts Curriculum Theory.* Fairmont, WV: Fairmont State College, 1980.
7. Harrison, G.B. "The Essence of School Technology" In: *School Technology Forum,* Working Paper No. 1

(H.M. Inspectorate). Published for the Standing Conference on Schools' Science and Technology (SCSST). Nottingham, Eng: The National Centre for School Technology (NCST), Trent Polytechnic, February 1973.

8. Mesthene, Emmanuel G. "Technology and Wisdom". In: *Technology and Social Change*. Indianapolis: Bobbs-Merrill, 1967.
9. Olson, Delmar W. *Industrial Arts and Technology.* Englewood Cliffs, N.J.: Prentice Hall, 1963.
10. Phillips, Kenneth "A Progression of Technology in Industrial Arts Education" In: *Technology Education:A Perspective on Implementation.* Reston, Virginia: International Technology Education Association, 1985.
11. Rapp, Friedrich. *Contributions to a Philosophy of Technology.* Dordrecht, Holland: D. Reidel Publishing Co., 1974.
12. Savage, Ernest and Sterry, Leonard "A Conceptual Framework for Technology Education," in progress. Technical Foundation of America.
13. Skolimowski, Henryk. "The Structure of Thinking in Technology" In:*Technology and Culture*, VII, No. 3 Chicago: University of Chicago Press, Summer 1966.
14. Todd, Ronald."Theory and Model Building in Industrial Arts and Technology," *Epsilon Pi Tau Journal*, Spring 1975.
15. Todd, Ronald. "Technology Education: An International Perspective" In: *Technology Education:: A Perspective on Implementation.* Reston, VA: International Technology Education Association, 1985.
16. Todd, Ronald." Science and Technology: Making the Links." Paris: UNESCO, in progress.
17. van der Velde, Jenne. "The Subject Technology in Dutch Education" In: PATT (Pupils' Attitude Toward Technology) Conference Report, p. 426. Eindhoven: Technische Universiteit, 1987.
18. Wiener, Norbert. *Cybernetics: Or Control and Communication in the Animal and the Machine.* New York: John Wiley and Sons, 1948.

Technology Education as a Component of Fundamental Education: A National Perspective

Barry E. Stern

U.S. Department of Education

Abstract: Technology education is a bright new hope in curriculum reform. It provides school children with important content and contextual information about technology while using successful teaching methods which emphasize integrated, holistic, multi-disciplinary, multi-sensory, hands-on learning. While the extension of this curriculum to more schools and students is desirable, like previous curricular reforms it will have little impact on student achievement if structural reforms to improve the productivity of the educational system are not implemented as well. Such structural reforms include the development of accountability systems with performance standards, school-based management, more choice of schools and providers of curricula within schools, alternative credentialing of teachers, using technology to increase productivity, longer and more flexible school days, more performance-based (as opposed to attendance-based) funding, and deregulation in exchange for improved performance. Federal resources should be used to effect both structural and curriculum change, since the implementation of curriculum change is unlikely to have much impact without eliminating structural barriers in the educational system that impede change.

Key words: curriculum reform, structural reform, productivity, technological literacy

Introduction

Today, as we approach the 21st century, the importance of technology literacy to the overall technological achievement and economic health of the U.S. and other industrialized countries cannot be overestimated. In each of these nations, the skills required of the work force and the technological sophistication required of the general population are increasing rapidly. Countries which fail to adapt to these changes will risk a decline in their international competitiveness and hence their standard of living.

In the United States, the quest for technological literacy parallels our concern for the health of our entire educational system. In March 1990, President Bush and the nation's 50 governors established six national education goals for the United States to reach by the year 2000. The six

NATO ASI Series, Vol. F 78
Integrating Advanced Technology into Technology Education
Edited by M. Hacker et al.

goals cover the entire spectrum of educational policy in the U.S. from preschool to adult education and address major problems in the country's educational system from low graduation rates to drugs in schools to poor achievement in math and science. If the U.S. is to achieve these goals, especially the goal on math and science achievement by the year 2000, technology education is likely to play an important role.

Responsibility for reaching this ambitious goal will be shared by many people and institutions, but the solutions are more profound than simply trying harder, or urging more cooperation and coordination among agencies. If we are serious about improving math and science achievement, and indeed, the overall educational performance of our students, we must explore different ways of teaching and organizing curricula.

Technology education is one of those ways, but it will not solve all our problems. Without other fundamental changes in the structure our educational system, technology education, like other curricular changes of the past, will have little overall impact on the achievement of our students.

The purpose of this paper is to examine the interaction between curriculum change and structural change, using technology education as an example. Data and impressions will be shared on the context of technological literacy and technology education in the United States, and the current role of the federal and state governments in advocating and supporting this curricular innovation. The paper then discusses how the curriculum model of change rarely (at least in the United States) fulfills its promise of significantly improving student achievement, particularly if it is used exclusively and implemented without the massive infusion of new resources. The paper concludes with how the adoption and performance of technology education can be enhanced through elimination or attenuation of certain structural barriers in the school system. One role of the federal government in overcoming these barriers will be examined as well.

First, permit me to offer some remarks to provide a context for this discussion.

The Office of Vocational and Adult Education (OVAE) is the principal office at the U.S. Department of Education with responsibility for vocational and technical education. As such, we administer the Carl D. Perkins Vocational and Applied Technology Education Act, the Adult Education Act, and many national discretionary programs. Altogether, OVAE currently administers programs costing $1.2 billion annually and influences how states and localities spend another $9-10 billion of their funds on vocational and adult education. We also administer the Technology Education Demonstration Program which was authorized by the 1988 Trade and Competitiveness Act. This is a very small program -- only $1 million has been appropriated so far -- yet it is the only federal discretionary grant program directly related to technology education. So our activity in supporting this model is very limited and very recent, and we duly acknowledge that the expertise in technology education is in the states and the local school systems which have implemented this curriculum.

In addition to the limited experience of the federal government with technology education, another context-setting observation which needs to be made at conferences like this one is the unique nature of the American educational system. What appears to be true in the U.S. is peculiar to our culture and not necessarily comparable to other countries. It should be remembered that the United States' system of education is very decentralized, and most of the decisions affecting educational policy are made at the state and local levels.

This paper will make three main points:

1. Technology education is the latest in a series of curriculum upgrades which have attempted to narrow the gap between what is taught and what society demands of its citizens. This curriculum model of change has a poor track record, particularly when large dollar amounts are not available to spread the curricular innovation.
2. There is a very real and potentially inhibiting dichotomy between "technology education" and "educational technology." This apparent dichotomy is a false one and probably does a disservice to the adoption of either. Moreover, unless technology educators come to appreciate "education" as a system, their vision of "systems education" and integrated curricula is unlikely to be realized.
3. One of the most important roles the U.S. government can play to enhance the adoption of technology education is to remove barriers that inhibit change in the educational system while providing incentives for curricular innovations that work. This role highlights the difference in what is sometimes called the "structural model of change" as opposed to the "curriculum model of change." In order for fundamental and beneficial change to occur, the federal government should focus more on the structural model than the curriculum model of change.

Technological Literacy in the U.S.

The U.S. historically has been a technologically oriented country. As a people, we surmise that our degree of technological success has been due to a combination of our free enterprise and market-driven economic system which encourages and rewards innovation; our democratic political system which allows for the freedom of thought and speech; our tradition of individualism which motivates people to invent and to be acknowledged by their peers; the independence of our universities and the investment we have made in postsecondary education, particularly our investment in basic research; and our wealth of natural resources and wilderness which gave us the wherewithal and motivation to tame or control our physical environment.

Curiously, few historians would cite our educational system as a principal reason for our degree of technological success. Indeed, one could argue that we are successful in this area despite our school system, which emphasizes theory over practice, principle over application, and performing operations rather than being able to use them to implement solutions to problems.

Perhaps the apparently limited role of U.S. education in this regard is due to the unclear relationship between technological success and the level of technology literacy in the general population. To be sure, it is important that the people who work with technology have some degree of technological literacy. But do all of us need it? Many become technologically literate in the course of living and working, but they did not get it in school. On the face of it, nevertheless, it would appear that higher levels of technology literacy in the general population would be a good thing.

In a world that is becoming more technologically complex, a society which is unable to raise or even maintain the levels of technology literacy of its citizens may not be able to make fundamental decisions about its future -- decisions, for example, on expanding the use of nuclear power, deregulating important industries like telecommunications, studying the problems of ozone depletion in upper atmosphere, or determining when to turn off life-support machines.

Furthermore, there is an unclear relationship between technology literacy and technology education. The relationship appears to be positive, *i.e.* students who study technology education ought to have a better understanding of mathematical and scientific principles, yet there is little if any data on the impact of technology education on outcome measures such as math and science scores or the ability to reason and solve problems.

Assuming that higher levels of technological literacy are important for a society to control rather than be controlled by technology, there is evidence to suggest that our growth in technological literacy has not kept pace with expanding technology and the success of other industrialized countries in understanding technology.

In the 1988 International Assessment of Educational Progress, 13-year-olds from the U.S. scored in the lowest group in mathematics and science proficiency. In mathematics, 40 percent of the Korean students demonstrated the ability to deal with complex concepts, compared with 9 percent of American students. Similarly 33 percent of the Korean students could apply intermediate scientific principles, compared to 7 percent of the American students [6:32]. While many of the tests include data on math and science achievement, there are no international comparisons of technology literacy. In math the U.S. finished behind students from Korea, Canada, Spain, the United Kingdom, and Ireland. In science the U.S. place second to last, ahead only of Ireland. Even more troubling is that, despite considerable investment in education, U.S. scores in relationship to other countries have been consistently low for the past 20 years.

Technology Education in the U.S.

Technology education as a field is only five to eight years old. It grew out of the prevocational industrial arts curriculum -- or "shop" -- which in most schools consisted of woods, metals, gas engines, and printing. As enrollments in vocational education declined during the 1980s because of a shift in emphasis (in most states) to classical academic subjects, industrial arts enrollments dropped also.

At the risk of oversimplifying how technology education came to be, one might surmise that the major push came from industrial arts educators who needed a marketable "product" in order to keep their jobs. Students and their parents were simply losing interest in the old shop courses, and states and local school boards, attempting to respond to the U.S. Education Department's *A Nation at Risk* study in 1983, were looking for elective courses to cut in order to make way for more course hour requirements in English, math, science, social studies, and foreign languages.

Rather than view technology education as a mere replacement of "shop" courses for the prevocational student, technology educators tend to align this curriculum more with the general preparation of all students. With the renewed national emphasis on math and science, and the growing realization that the traditional methods of teaching these subjects do not work for a large proportion of students, the time would appear to be right for technology education.

Definitions

One of the difficulties facing technology education in the U.S. is its lack of a clear and precise definition. There is generally agreement on some of the key elements. These usually include the integration of academic disciplines into world of work "systems," experiential or holistic learning, creative application of knowledge to solve problems, and the teaching of basic concepts using concrete examples.

Although there may be general acceptance of the common principles and objectives of technology education, there is no legal definition. Different laws have different definitions. The 1988 Omnibus Trade and Competitiveness Act defined technology education as: "A comprehensive educational process designed to develop a population that is knowledgeable about technology, its evolution, systems techniques, utilization in industry and other fields, and social and cultural significance" [9].

The Carl D. Perkins Applied and Vocational Technology Act uses a narrower definition: "An applied discipline designed to promote technological literacy which provides knowledge and understanding of the impacts of technology including its organizations, techniques, tools, and skills to solve practical problems and extends human capabilities in areas such as construction, manufacturing, communications, transportation, power and energy" [2].

This lack of a clear definition has hindered federal efforts to identify and encourage exemplary technology education programs. Despite such broad definitions in legislation, however, many states have begun to develop curriculum frameworks for technology education.

Local Examples of Technology Education

In Ann Arbor, Michigan, two technology teachers at Forsythe Intermediate School are implementing a total curriculum change. They have changed their program from courses in hard materials, soft materials, graphics and mechanical drawing, and power technology to a "technology education" curriculum which features courses in communication, construction, manufacturing, and transportation. This program is a first for the state of Michigan, and although a substantial number of Michigan schools and instructors have incorporated some technology education modules in their programs, a complete technology education curriculum has not surfaced.

In Bellevue High School in near Seattle, Washington, class names have been changed from "industrial arts" to "technology" and the emphasis now is on creativity and students working together to find solutions to problems. The goal of the Bellevue High technology education program is to prepare students for collegiate and/or advanced vocational/technical programs and equip them with college-transferable technological skills that will also be useful to them as citizens, voters, and consumers [5:32]. The objectives of this course are that, as a result of technology education, the students will experience the practical application of basic scientific and mathematical principles; make decisions regarding technology related careers; gain an in-depth understanding of and appreciation for technology in our society and culture; develop basic skills in the proper use of tools, machines, materials, and processes; and solve problems involving the tools, materials, processes, and products of industry technology.

At John Hanson Middle School in Charles County, Maryland, the technology education program is based on a team-teaching approach. Eighth grade students spend all year in this program, which begins with a 4-6 week hands-on machine/tool orientation designed to develop the skills necessary to function safely and competently in the laboratory during the rest of the year. After this period the students undertake an in-depth study of technology. Since they can choose the topics to study after the internal orientation, the students participate in the program with a high degree of motivation.

The "Triad Technology Education Cluster" at Triad High School in Troy, Illinois, was established in 1980 as an innovative technology education curriculum. Areas of technology were "clustered" or organized in the course format to help students understand how they were related to one another. Initial development of the program was based on a list of general objectives developed by Triad staff. Different sections of the course are taught by four teachers. Courses are offered on an elective basis and are divided into three different levels of study on a sequential basis. In Level One the student enters the program by taking a course to introduce and explore various technological areas. Level Two students continue to develop their skills by choosing from various skill development courses. Level Three students may study more advanced technologies as well as the societal consequences of these technologies.

Although many of these new programs are being implemented on a small scale, there are examples of entire states using the type of "systems" curricula prescribed in technology education. In the state of Kansas, for example, the "Exploration in Technology" program that was started at Pittsburgh Middle School now serves as a model for the entire state. This self-directed, self-paced program places the responsibility for learning directly on the students and provides them with experiences in 20 different technology modules.

Several successful programs such as these are operating around the country. Many more are expanding, and new ones are being started every year. Unfortunately, however, technology literacy standards against which to measure curricula have not been developed, and we know little about the impact of these advanced curricula on outcome measures such as test scores and graduation rates.

Federal Role in Technology Education

The federal government is also involved in supporting technology education, although to this point its involvement has been limited. Under the Technology Education Demonstration Program mentioned earlier, the Department of Education is funding projects in Alaska, West Virginia, Wisconsin, Connecticut, and Oklahoma. The purpose of this program as defined by the legislation is to "assist local educational agencies and institutions in developing a technologically literate population through instructional programs in technology education" [9]. The legislation further states that course content must be based on an organized set of concepts, processes, and systems that is uniquely technological and relevant to the changing needs of the workplace and the impact of technology on people, the environment, and culture. The Congress also mandated that instructional content be drawn from established technology education courses in one or more of the following areas: communication, construction, manufacturing, and transportation.

The University of Alaska/Anchorage School District project called "Blueprint for Literacy in Technology" (BLT) involves two school districts and enrolls approximately half the students in Alaska. BLT has been developed as part of a public school/university collaboration led by two eminent national leaders who are recognized advocates of technology education as well as lead designers of the curriculum used in this project.

The goals of the West Virginia University Appalachian Technology Education Consortium are to contribute toward the development of students who are knowledgeable about technology, to enhance the capabilities of technology teachers, to contribute to the integration of math, science, and technology education curricula in secondary schools, and to disseminate results of the project. More than 136 educators and business/industry representatives will be directly involved in the regional Consortium. Most will receive special training. At least 2,000 secondary students will be involved.

At the Milwaukee Area Technical College, 100 minority students new to technology education will be enrolled in secondary technical education programs leading to associate degrees. Secondary schools and technical colleges will encourage underachieving, disenfranchised, at-risk youth to enter technical education and develop lifelong learning habits, problem-solving skills, self-confidence and positive work habits. However, because less than $1 million was appropriated for this program, the Department established a competitive priority which gave preference to applications that emphasized development of teacher capability, materials, and teacher workshops for interfacing math, science, and technology education.

The Oklahoma State University Tech-Ed demonstration project will use a multidisciplinary approach to develop teacher capabilities to integrate math, science, and technology education.

These projects are likely to be effective. Unfortunately, however, the extremely broad definition in the legislation and the fact that no regulations were written for this program resulted in many proposals that were not clearly focused.

Technology Education: The Latest Example of the Curriculum Model of Change

Parallels to Previous Curricular Innovations

Technology education is the latest example of changing schools to reflect what societies do and value. There are striking parallels between technology education today and previous curriculum innovations, for example, career education of the 1970s. Both curricula give considerable attention to the definition -- what it is, how you would know when you have it, and how you would measure its success. Proponents of career education and technology education continue to debate vigorously whether the respective curricula should be a separate course of study or infused into traditional curricula. Either scenario creates "curriculum squeeze" -- teachers ask what will be pushed out of curriculum to make room for this new content. Of course, many teachers won't change their courses until they get some pretty clear signals from the local school board that this new content is desired.

Along with curriculum squeeze is a perceived need to upgrade textbooks and to train or retrain teachers to deliver this new curriculum. Like career education, technology education emphasizes a progression from elementary to middle to high school: awareness in elementary school, exploration in middle school, preparation for work or higher education in high school, and revamped teacher preparation in colleges.

Both career education and technology education rediscovered John Dewey, who emphasized "learning by doing" and experiential, integrated education. Almost all curriculum innovations like to reinvent "good teaching," which progresses from the known to the unknown, from the general to the specific, from the simple to the complex, from ideas to fruition, and which provides the tools and analytic skills with which to inquire and solve problems. Good teaching has always taught students how to think and question basic assumptions and purported facts. And good teaching has always tried to relate content to the context in which it will be applied.

American vocational education has also rediscovered the concepts of disciplinary integration and skill progression. The new Carl Perkins Vocational and Applied Technology Education Act calls for both academic/vocational integration and a relatively new curriculum concept called "tech-prep." These concepts are related, but there is a subtle difference. Academic/vocational integration occurs when teachers of academic and vocational courses at the secondary level get together to develop new curricula that integrate theory with practice [3]. Tech-prep occurs when high school and community college teachers get together to develop curricula that carefully sequence the attainment of competencies from 11th grade in high school through two years of community college, ending with an Associate of Arts Degree or certification in a technical field [10].

Many practitioners in the field have enthusiastically embraced these concepts; in fact, much is already going on. Reauthorization of the Perkins Act is perhaps a first step by the federal government to confirm this trend toward disciplinary integration and "hands-on, minds-on" education. Future legislation may attempt to encourage integration not only for vocational education but for academic subjects as well. The U.S. needs to do a better job in implementing more holistic, integrated education. Technology education may well be the harbinger of the educational system of the future, since it coalesces the traditional disciplines around systems (transportation, communication, manufacturing, power and energy, food, space, etc.), employs proven teaching methods, mobilizes students to inquire and learn on their own, and teaches students the basic concepts of flow, conversion, storage, and risk/benefit.

Limitations of the Curriculum Model of Change

However excellent a concept technology education may be, it is still part of the curriculum catch-up model of change, and as such is limited in its capabilities. In theory, the curriculum model of change is supposed to narrow the gap between what is taught and what society demands. Many believe that if we can get that gap down to a decade, we would be doing well. The big question is "if."

Unfortunately, despite all the reforms and curriculum innovations we have seen in the U.S. over the last several years, there is scarce evidence that the lag or gap between school and society is closing, whether you use test scores, dropout rates, employer satisfaction with new hires, or any other measure. Rather than narrow the gap, our efforts seem directed toward preventing it from widening, or not losing further ground. The curriculum model of change has not worked either in terms of aggregate results or in terms of getting widespread adoption of the new curriculum.

Despite this record, there are those in the technology education movement who would say, "Schools must become proactive agencies in dealing with the study of technology and its implications for life in the 21st century." I don't know of any country in which public schools have been proactive -- out ahead of the curve. Certainly, there is no chance of that if the curriculum model of change is used. The only possibility of schools being proactive -- agents of change --is if they become self-renewing organizations that continually remove barriers to and manipulate incentives for positive action.

There are several inherent limitations to the curriculum model of change. For example, the curriculum model of change does not address the reality of 50-minute time traps in which to conduct a lesson. Nor does it address the issue of choice vs. the monopoly of the public school system to deliver with public funds a particular course or curriculum. Nor does it address the inexperience of teachers in systems other than education, the problem of teacher credentialing, the lack of incentives for teachers to learn and employ new curricula, the use of technology to increase the productivity of school systems, and the lack of performance measures and standards with which to hold schools accountable.

Unless barriers such as these are addressed, curriculum change -- the things that students should be exposed to because it is good for them -- will be slow to proceed.

Eliminating Barriers to Curriculum Reform

This brings me to the federal role regarding curriculum change in general and the more widespread adoption of technology education in particular. There are two central questions which must be addressed:

1. Should the federal government favor technology education over other subjects or disciplines?
2. Should the federal government provide technical assistance and training to encourage more widespread adoption of technology education?

Rather than answer such questions with a simple "yes" or "no" -- or "it depends" -- it is helpful to look at how government works best. To promote adoption of innovations, there is evidence that manipulating "barriers and incentives" can be effective government policy [11:47]. Government works best at the extreme ends of the innovation process: supporting basic R&D and removing barriers to adoption of desirable innovations. Markets and private enterprise are the best engines we have found for the actual creation of innovation [11:48].

What are the barriers that the federal government could help eliminate and the incentives that it might help promote to encourage more widespread adoption of technology education?

Accountability and Performance Standards with School-Based Management

Currently, in most schools there are no reasons for school personnel to adopt modern curricula like technology education. There are no consequences for not adopting, or for that matter, for not getting students to achieve. Some local school systems, however, and some states (like California) have adopted the concept of school-site report cards so a parent or a citizen can know which schools are performing best or progressing most.

The federal government is nurturing the trend toward more accountability through the establishment of national goals and measurement systems to keep track of progress in meeting them. President Bush and the nation's governors have agreed to an ambitious set of education goals for the country to reach by the year 2000. Progress is currently being made in setting up assessment systems to measure achievement toward these national goals. A major structural issue in U.S. education is to strike a balance between achieving national and state educational goals and decentralizing the management of schools down to the school site level. Stated another way, how do you establish national performance standards while decentralizing authority for achieving those standards?

The trend toward more accountability and outcomes-based education should be the friend of technology education and other curriculum innovations. When the innovator can demonstrate that the new curriculum helps students achieve performance standards better, faster, with more joy, the world, as they say, will beat a path to her door.

However, accountability with decentralized management poses yet another challenge: it is much more difficult to get widespread adoption with decentralized authority. It may not be enough, therefore, to convince the superintendent, but the principal, faculty, and parents as well. If technology education can show better test results and provide evidence of retaining students in school, schools will be more likely to adopt it.

Choice of Schools and Providers of Curricula

Choice is a currently a hot political issue in the U.S. There are some fundamental benefits that choice in public schools can provide. Choice helps technology education by allowing parents and students to select schools which have a strong technology education curriculum as reflected by high or improving scores on school achievement tests. Many states have introduced various choice plans in which students and parents can select their schools within certain parameters (*e.g.*, any public school in a metropolitan area).

But selecting among schools is not the only type of choice. In addition, there should be an opportunity to choose among providers within a curricular area. For example, high school students ought to be able to take vocational training in high schools, area vocational-technical institutes, or community colleges. In the future, private entities such as a proprietary schools or companies that wish to offer vocational training or apprenticeships to public school students should be allowed to compete for training contracts as well.

Choice within a school, in other words, should not necessarily be limited to instruction provided by the school. Schools should be able to widen their choices by contracting out courses that could be offered more appropriately by non-public school entities. An example would be a company that wishes to offer a vocational course in a field where equipment is expensive yet becomes technologically obsolete in a short time. Another example of needlessly limited choice is not having access to a technology education program because of oversubscription to the school's program. If a program is popular, then an instructor or team of instructors should be able to teach more sections. This allows outstanding teachers to reach more students while retraining or upgrading the skills of other teachers.

More Technology-Intensive and Less Labor-Intensive Education

Technology education and educational technology are admittedly different concepts, but there will be no technology education without more technology at the disposal of the teacher. Interactive video disc technology, for example, makes it possible not only to provide effective individualized instruction, but also to diagnose individual learning needs inexpensively. This would seem to be particularly important in technology education, where students are engaged frequently in team problem-solving but lack the skills to be fully contributing members of the team. Educational technology can help teachers see to it that basic skills in different disciplines do not get lost as they are trying to integrate and apply disciplinary knowledge to solve systems problems in the real world.

Is technology itself a barrier? Technology per se is not the barrier. Instead, it is our attitude about using technology in schools that is a barrier. Our attitudinal problem in the U.S. is that educators are reluctant to move toward a system that is more technology-intensive and less labor-intensive. There are several ramifications to this problem.

In the area of technical systems, the latest generation of computer technology is vastly extending the practical application of artificial intelligence in general and expert systems in particular. The potential to automate most of the didactic functions of schooling is about to be realized. Advancement of learning technology will blur if not obliterate the authoritarian distinction between teacher and learner. Teachers who feel threatened that their work is being taken over by automation would obviously constitute a barrier to innovation.

Too many educators want technology as an add-on to the educational enterprise, which, of course, drives up the cost. Most are unwilling to use technology to reduce instructional costs or increase the productivity of the educational system. Evidence of this is the continual reappearance of the class size issue in collective bargaining. This predominant attitude on the part of educators has resulted in a stagnation if not a lowering of educational productivity in the U.S. while other industries, in order to survive, have had to increase their productivity via technology and improved management systems.

In the next several years, it is unlikely that there will be in the U.S. a large increase in the public investment per student. Under these conditions, teachers have to decide whether they want additional resources spent on: (1) smaller classes (more teachers) with little left to invest in technology, which will result in continued low pay for teachers relative to other professions; or (2) more technology and perhaps larger classes (to make it possible to pay for the technology), which will result in more highly paid teachers with more professional discretion to organize courses and use both technology and human resources to achieve desired results. Those in technology education have the opportunity to help shape this debate. But we must be careful not to perpetuate the myth that technology education and education technology are different things. Without a more technology-intensive educational system in which teachers control more technology to increase their productivity, the truly great things that technology education has to offer will continue to be provided to a small minority of students.

In short, the education industry has to make the decision to begin to substitute technology for people. Along with that is a change in how we utilize our human resources. The factory model of putting a teacher in front of a class of students for 50 minutes at a time is no longer adequate and certainly not very effective considering the technological resources at our disposal. The role of technology in the educational system has to be appreciated by technology educators just as they would consider technology's role in transportation, communications, or energy systems. Not to do so by continuing to rely on the resource-guzzling curriculum model of change will keep technology education a marginal activity in most school districts and classrooms.

Alternative Certification of Teachers and Administrators

Despite low pay relative to other professions, there are highly qualified people in industry who would like to teach, but not badly enough to meet what they perceive to be onerous teacher certification requirements. Archaic certification requirements in the U.S. make it extremely difficult if not impossible to recruit these people into teaching. This is a real issue for technology education, since most of the experts on the kinds of systems included in technology education (*e.g.*, communications, manufacturing) do not work in schools.

Several states are adopting alternative certification requirements in order to facilitate the recruitment of talented individuals who would like to share their craft and knowledge with students. Even the major teachers' unions, the National Education Association and the American Federation of Teachers, have endorsed this concept so long as there are safeguards to verify the competence of individuals who wish to enter the teaching profession through alternative routes.

We should also explore alternative ways of certifying school administrators. Why shouldn't an extremely talented manager in another industry be allowed to be a principal without having to go through an extensive credentialing process? Why shouldn't teachers be able to select their principal from their own ranks? As long as accountability systems are in place to assure performance, these alternative certification procedures are unlikely to do any harm.

Perhaps more than alternative certification is the desire by teachers, particularly teachers of technology education, to experience first hand the world of work systems they are teaching about. In addition to teacher-industry exchanges are teacher sabbaticals in business. Such a program was set up in the Seattle Public Schools with the help of Partnerships in Public Education (PIPE), Seattle's school-business partnership program. Teachers brought with them their half-pay sabbatical stipend, and the company which hired the teacher matched the stipend. The teacher in effect worked for the private business during the sabbatical period.

Technology education would be well-served by this kind of program, although I suspect the Seattle model would be extremely expensive if the same financial arrangements were employed. Sabbaticals should be much more the norm than the exception in education, and teachers should expect significant outside-of-school experiences -- not just summer workshops -- to be part of their careers as educators.

Other Structural Barriers to Change

Flexible/longer school day -- This option must be put into use if students are to have the time to work on team projects and engage in problem solving and self-directed study to acquire the skills to be able to carry out team projects. Technology education would appear to require such flexibility.

Regulations -- State and federal regulations often get in the way of curriculum innovations which seek to integrate disciplines. Some states (*e.g.*, Washington) and local school districts (*e.g.*, Marin County, California) are experimenting with ways to relax regulatory requirements in exchange for getting results. The term we use in the federal government for this tradeoff of "rules for results" is "EDFLEX." Removal of regulations barriers would seem to be particularly appropriate for technology education. The field should benefit from a decision made at the "Education Summit" by President Bush and the nation's governors to work together to minimize regulations barriers.

Performance-based (as opposed to attendance-based) *funding* -- This is a major structural reform that would favor most any curricular reform that works. Imagine a teacher who is given a budget to ensure that a class of 25 students learns skills in math, science, English, and social studies through a unit on communications that s/he was assigned to teach. The teacher could decide how much of the unit to teach personally and how much to contract out to other teachers or private companies. Since part of that teacher's pay, and part of the state's reimbursement to his/her school district, would depend on the students' aggregate performance, there would be a powerful incentive to deliver the strongest program possible. Teachers would then have to marshal all the human and technological resources possible to encourage student learning and to assess their progress. This "professional practice" model -- where teachers either alone or in professional groups become contractors who sell their services to the system -- is already being tried [11:110], but it is not yet part of mainstream American education.

Although the U.S. is far away from this model at present, performance funding would appear on the surface to be a powerful incentive. The carrot is both more pay and more professional discretion over resources; the stick, of course, is less pay and perhaps less discretion in the future for lack of success. This model may well be the one to drive technology education in the 21st century because it requires that teachers use educational technology to become more productive.

Federal Role in Fostering Structural and Curriculum Change

During the decade of the 1980s, the investment by U.S. taxpayers in each public school student increased by 26 percent in inflation-adjusted dollars. Yet little, if any, improvement was noted in test scores, dropout rates, or any other criterion of success. Most of the increased investment during this period supported the curriculum model of change -- more course requirements, more resources for special populations, and more teachers and other school personnel.

As a result of these discouraging results, the discussion is shifting in the 1990s from "reform" to "restructuring." I suspect that federal resources will strike a balance between supporting curriculum change and structural change. Since large budget increases are unlikely, it may well be that the use of resources to foster structural change will go further toward implementing technology education (or any other desirable curriculum) than will the mere direction of federal resources to train teachers. Federal technical assistance and training would constitute yet another "add-on" innovation with external support ("soft money"). Yet without internal structural change -- especially commitment of resources in the local school system's core budget -- the innovation is unlikely to be successful [11:56].

Another lesson from government's experience in trying to foster the widespread adoption of a successful curricular innovation is that informal systems (in the U.S. at least) seem to work as well as if not better than government-planned formal systems. For example, while government-promoted attempts to promulgate computer literacy through formal education have been a fiasco, millions of Americans have rapidly acquired potent mastery of personal computer technology through users groups and thousands of other personal networks [11:104].

In the United States, government probably should foster the adoption of curriculum innovations by building on our capacity to associate and network rather than attempt to set up yet another technical assistance and teacher re-training program. Maybe we should set up an information system that would help practitioners answer their questions about technology education, or for that matter any curriculum. Teams of practitioners could be established around the country to frame the questions. Teams of researchers in the same curricular areas could be convened to answer them. And software experts could be hired to develop strategies to access the questions and answers through a computerized "expert system."

The U.S. Education Department is already moving in the direction of making the informal system of networking and association work better. We have dozens of public and private educational clearinghouses, research and development centers, and educational laboratories, and we have a cornucopia of meetings, conferences, symposia, and colloquia to share findings from educational research. These informal systems, however, are only productive to the extent that the information being networked is accessible and high quality.

Improving our systems of disseminating information is becoming a major activity of the U.S. Department of Education. The technology has been there for some time but is underutilized. Now that information overload is occurring, more of my colleagues at the Department are perceiving the need to separate the wheat from the chaff and to get the best wheat, *i.e.*, information, into the hands of the breadmakers, *i.e.*, the practitioners. Perhaps as much as anything we will do in the next few years, this will help the cause of technology education and other curricular reforms that work.

Summary

Technology education is a bright new hope in curriculum reform. It provides school children with important content and contextual information about technology, while using successful teaching methods which emphasize integrated, holistic, multi-disciplinary, multi-sensory, hands-on learning. While the extension of this curriculum to more schools and students is desirable, like previous curricular reforms it will have little impact on student achievement if structural reforms to improve the productivity of the educational system are not implemented as well. Such structural reforms include the development of accountability systems with performance standards, school-based management, more choice of schools and providers of curricula within schools, alternative credentialing of teachers, using technology to increase productivity, longer and more flexible school days, more performance-based (as opposed to attendance-based) funding, and deregulation in exchange for improved performance. Federal resources should be used to effect both structural and curriculum change, since implementation of curriculum change is unlikely to have much impact without eliminating structural barriers in the educational system that impede change.

References

1. American Association for the Advancement of Science. *Science for All Americans.* AAAS, 1989.
2. *Carl D. Perkins Vocational and Applied Technology Education Act Amendments of 1990.*
3. Grubb, W. Norton, Davis, Gary, and Lum, Jeannie. *The Cunning Hand, the Cultured Mind: Models for Integrating Vocational and Academic Education.* Berkeley, Ca: National Center for Research in Vocational Education, 1990.
4. International Technology Education Association. *Technology Education: A Perspective on Implementation.* ITEA, 1985.
5. International Technology Education Association. *Technology Education in Action: Outstanding Programs.* ITEA, 1989.
6. National Center for Education Statistics. *1990 Education Indicators.* Washington, DC: U.S. Department of Education, 1990.
7. *National Education Goals: A Report to the Nation's Governors.* Washington, DC: The White House, 1990.
8. National Governor's Association. *Educating America: State Strategies for Achieving the National Education Goals.* NGA, 1990.
9. *Omnibus Trade and Competitiveness Act of 1988*, Public Law 100-418.
10. Parnell, Dale. *The Neglected Majority.* Washington, DC: The Community College Press, 1985.
11. Perelman, Lewis J. "Technology and Transformation of Schools: Public Education Strategies for a More Competitive Nation," Version 1.1. March 1987.
12. Stern, Barry. "New Directions for Vocational Education: Implications for Data Programs, Career Guidance and Counseling." Speech to the Thirteenth Annual National State Occupational Information Coordinating Committee, August 1990.
13. Technology Education Advisory Council. *A National Imperative.* ITEA, 1988.

III. Technology Education and Other Curriculum Areas

Technology has relationships to other areas of human activity. As was already recognized in the presentations and discussions in the first session, advanced technology in particular is related to science. This relationship was explored further here as well as relationships to other curriculum areas.

Mr. Giles Job (Staffordshire LEA, UK) stated that models for science and technology are often simplifying the relationship between these two. He showed some of the conceptual and methodological differences between them. On the school level this implies that pupils should realize that science and technology are different but related.

Dr. Marc de Vries (Pedagogical Technological College, the Netherlands) emphasized the complexity of the science-technology relationship. Historically they have become partners, which strongly influenced technology's methodology. Therefore in education this relationship should be made clear to pupils.

In one discussion group it was stated that technology education should be seen as a centerpiece to all curriculum areas, because technology is interdisciplinary in nature and affects all aspects of human life. Technology education is not the matter of one teacher, but it has to do with the whole school climate. One of the other groups formulated a *caveat:* there is a danger to over simplifying the science-technology relationship that leads to ideas such as the possibility of teaching scientific concepts through technology, for example, which can be questioned. The integration of advanced technologies into technology education increased the need for clarifying the science-technology relationship to students because it is this relationship that, to a large extent, determines the nature of advanced technology.

Various options for showing the science-technology relationship in education were mentioned: integrating science and technology into one subject, doing cross-curricular projects in which both science and technology play a role, integrating elements of technology into all subjects. All have their pros and cons, and age level is one of the variables that determines which option is most desirable.

The Relationship Between Science and Technology in the School Entitlement Curriculum

Giles C. Job

Staffordshire Education Department, UK

Abstract: The relationship between science and technology in the school entitlement curriculum is bedevilled by simplistic stereotypes. Scientific and technological awareness should be complemented by development of scientific and technological capability. Effective relationship in the curriculum must take account of the distinctive ways of knowing in scientific and technological activity and of children's ways of learning.

Key words: science, curriculum, entitlement curriculum, national curriculum, curriculum model, technological awareness, investigation, paradigm, task-action-capability, propositional knowledge, transactional knowledge

Introduction

This paper discusses the relationship between science and technology as components of the common or core curriculum for children of compulsory school age. It focuses particularly upon the English situation, where science and technology are now tightly specified compulsory subjects within the National Curriculum for all children between the ages of 5 and 16 [10,11]. The natures and origins of the paradigms of science and technology underlying their curriculum realization are discussed, and the implications for further curriculum development and research are considered.

In most Western societies, the inclusion of technology as a component in the core or common curriculum of all children is a relatively recent phenomenon, whereas the case for science, admittedly with great variations in the scope of the common science curriculum, is well established [9,19]. In England the recent establishment of technology as a foundation subject within the National Curriculum follows only about 15 years after the first really significant national policy proposals that all children are entitled to education within a technological "area of experience" [9,19].

NATO ASI Series, Vol. F 78
Integrating Advanced Technology into Technology Education
Edited by M. Hacker et al.

Relationships between Science and Technology

The Need for a Rationale

The close relationship between science and technology is a truism that is given popular and academic lip-service, but the nature of this relationship in its significance for the content and pedagogy of the science and technology curricula is remarkably rarely clearly articulated. In the absence of such articulation, it is perhaps not surprising that curriculum developers and teachers tend to revert to simplistic models of science and technology.

Popular Perceptions

Certainly in common public perception, there is confusion and uncertainty about the nature of and relationship between science and technology. One common usage does not really attempt to differentiate between the two. "Scienceandtechnology" becomes in effect a single entity; the roles of scientist and technologist are interchangeable; they both deal abstrusely with materials and devices of great complexity and potency; they both wear white coats. There is evidence that this view is widespread internationally and affects the attitudes that pupils bring to their education [19].

Another public perception confuses and largely identifies the roles of the technologist, the engineer, and the technician; they all make and mend things involving machinery; if it gets dirty they exchange their white coats for blue overalls. This caricature is still unfortunately prevalent in the UK, and is obviously related to the depressingly low public status afforded to engineers and technologists.

If these two views have anything in common about the natures of science and technology, it is probably conveyed in the words of a manufacturer's advertisement for a range of household appliances: technology is "the appliance of science." An obvious aim for technology education is to redress these misperceptions and to raise the status of the profession of technologist [20]; but to succeed in such an aim requires close attention to the explicit and implicit messages about technology and its relationships to other areas conveyed to pupils by the curriculum.

Categorizing "Science" and "Technology"

In exploring this relationship further, it is probably more fruitful to consider a number of distinct categories of the uses of the words "science" and "technology" rather than to indulge further in the most time-consuming of English educational games of the last few years--producing a single adequate definition for technology. Particularly relevant to the curriculum debate are considerations of science and technology as:

- academic disciplines
- social and economic areas of activity
- cultures or world-views
- curriculum areas of activity.

As *academic disciplines,* science and technology, or more exactly the sciences and technologies, are well established, and have their own professional structures, internal logics, and arguments. Expertise and skill in specific technologies will require some skill and expertise in specific aspects of science, and to some extent vice versa. But the logic of these interrelationships in specialist and professional education and training does not automatically hold in the field of school curriculum entitlement.

In considering science and technology as social and economic activities, it is constantly startling to realize how limited and recent is any overt and self-conscious interaction between the two. Until very recently, technological activity has produced complex, sophisticated, and beautiful artifacts and systems in which the overt application of scientific theory or of systematic experimental enquiry has had absolutely no part. Even today, the involvement of science and scientists in large-scale technological activity is often at best sporadic. Primo Levi, for example, writes tellingly and at times hilariously of the vicissitudes of a chemist in Italy's developing chemical industry in the mid-twentieth century, and of the uneasy and slow merging of a strong, practical, and pragmatic craft tradition with a more recent but equally strong tradition of systematic, empirical enquiry based on a structure of simple but abstract theory [18].

Even in the newer "high" technologies, which have had from their inception a close relationship with emerging scientific disciplines, the relationships are very complex from the simple and rather quixotic models of the creative technologist exploiting the findings of a precursive scientific principle; or vice versa of the scientist carrying out the research necessary to surmount an obstacle on the way to a solving a technological problem. And the more complex the technological activity, the more difficult it may become to separate its science and technology dimensions. It can be argued that in a very real sense the exotic subatomic particles "discovered" in the last 20 years are as much technological products of the complex systems that produce and detect them as they are physical entities. There are no simple paradigms here on which to model science and technology in the school curriculum.

Also relevant to this curriculum debate are the *cultures of science and technology*--the values and attitudes associated with scientists and technologists at work, leisure, and thought. Among the contrasting aspects of these two cultures, the following can be regarded as particularly significant to the curriculum debate (given, of course, the dose of salt essential to all such cultural generalization) [Sparkes, R. in 8]:

Scientific	Technological
analytic	synthetic
specialist	interdisciplinary
knowledge-driven	problem-driven
cooperative	competitive
free exchange	commercial secrecy

It may be argued that the fundamental distinction underlying these and other contrasts is one of motivation and world-view. The world of technology has to do with the urge and the need to change and improve the environment, to extend human capability. The world of science has by contrast to do with the urge to know and understand the environment. The realization that these two urges and their outcomes can be systematically harnessed together is perhaps the major human intellectual and evolutionary achievement of the 20th century [15,16].

Technology as the "Third Culture"

The view of technology as the "third culture" to sit alongside the arts and science is perhaps to strain the English word "technology," but that is the fault of the language and the influences that formed it, not of the concept. The single English word has to do service for a number in other European languages that bear these loads better--for example, *Technik* and *Technologie* in German. Certainly, the lack of a suitably loaded and acceptable word has hindered the struggle to establish this area of experience in the core of the English school curriculum. In this context, it is interesting if not helpful to speculate upon the fact that the technological basis of the industrial revolution of the 18th and 19th centuries grew from craft and manufacturing traditions that expressed themselves predominantly in the monosyllabic Anglo-Saxon elements of the English language--the language of make, do, fix, mend, knack. By contrast, English scientific development from the 17th to the 19th centuries has mainly been associated with the formally educated and leisured classes and has been expressed in the polysyllabic Graeco-Latin elements of the language. Maybe technology should have been a four-letter word....

Science and Technology as Curriculum Activities

All these aspects must come to bear on decisions, conscious or unconscious, about the role and relationship of science and technology as components of a curriculum to which all children are entitled. Which aspects of these subjects is it desirable, appropriate, and sensible to give all children experience of? A crude but serviceable curriculum taxonomy that is particularly helpful for a consideration of science and technology is as follows:

- knowledge of the subject - "know about"
- knowledge in the subject - "know that"
- skills in the subject - "know how"
- attitudes through the subject - "know why"

The remainder of this paper uses these categories to consider recent curriculum developments in science and technology, and to highlight some aspects that warrant further consideration.

Technology as a Cross-curriculum Activity

An immediate complication arises with the word "subject." Over the last decade there has been in England a vigorous movement that argues for technology to be treated not as a conventional school subject or group of subjects but as a distinct style of learning and knowing that should be developed where appropriate through many or most conventional timetabled subjects--"technology across the curriculum" [7,19]. This movement has certainly brought to the fore of educational debate a serious consideration of the role of technology as the "third culture" in the English educational system and has stimulated some influential subject-teaching associations to define their attitudes towards and claims upon technology [1]. It is premature to judge the long-term efficacy of this approach to technology education. If curriculum coherence is to be achieved, major demands are made on a school's abilities to plan and coordinate the whole curriculum--a rock on which earlier "across the curriculum" movements in, for example, language and mathematics have largely foundered. The definition within the National Curriculum of technology as a foundation subject [11], with the requirement that a detailed assessment be made of each child's achievements in the subject at ages 7, 11, 14, and 16, further complicates this cross-curricular approach to technology.

Awareness of Science and Technology

In the current English context, there is general agreement that it is neither efficient nor sensible to begin the systematic specialist training of children in science or technology within the common, compulsory curriculum. There is less general agreement as to when some degree of specialization with its tacit acceptance of the professional training structures of the discipline should commence; but in practice any significant specialization before the age of 16 is now very limited. There is also general agreement between educationalists and politicians that science and technology are such potent features of the "technological world in which we live" that the common curriculum should aim to give young people an accurate and thoughtful awareness of the nature and role of science and technology in society. Such awareness, it is argued, will promote informed career choices, contribute to informed social and political decisions, and help people to feel comfortable in a "technological world."

Curriculum Models and Stereotypes

This demand puts on to curriculum developers the responsibility to ensure that curricula do not reinforce some of the potent oversimplistic stereotypes of science and technology discussed above, made more acute by the fact that very few teachers have ever themselves been practicing scientists or technologists, and very few science and technology teachers have any significant training in history, sociology, or epistemology. The requirement for an accurate awareness element of science and technology courses has to be seen in the long-standing English tradition that the essence of a school subject is the teaching of the subject, from which learning about the subject should emerge [19]. Courses that, for example, directly teach pupils about technology with relatively little emphasis on developing their subject skills are almost unheard of in England, unlike the USA, for example, where such courses not only are acceptable and well resourced, but also have as a consequence a well-developed theory and methodology.

Any course that attempts to give students a realistic awareness of the relationship between science and technology is faced by the inevitably limited experience and understanding of students, perhaps particularly in science. Two distinct approaches seem to be common. One is to relate some aspect of doable and understandable "school science" to a simple explanation of a feature of a technological artifact or system--*e.g.*, using simple practical and theory work on Bernouilli's principle to "explain" the movement of a wind generator rotor. This may root the scientific principle in a real, high-tech context but does nothing to illuminate the interplay between scientific and technological principles needed to make the generator a practical possibility. Another common approach is to extrapolate from a simple school-based application of a scientific principle to a large-scale, real-world analogue--*e.g.*, from a laboratory working model telegraph to satellite communication. Here the technological principle may be illustrated, but the science involved in the real-world example is irrelevant to its school model.

Promoting Scientific and Technological Awareness

Despite these shortcomings, there is nevertheless some structural progress in promoting consistent and accurate treatment of science and technology relationships in the English National Curriculum. All GCSE examinations in science (the examinations taken at age 16 by most students) are required to include assessment of the "social, economic and technological applications and implications" of the subject [23]. The National Curriculum for science [10] contains frequent reference to the role of science in technology, and also includes an attainment target (major objective) for 11-16 year olds on the nature of science--a thoughtfully structured and non-simplistic if challenging introduction to the history and philosophy of science. Such overt self-consciousness or self-reference is unfortunately relatively absent from the National Curriculum requirements for technology. Without explicit and thoughtful attention to the teaching of scientific and technological awareness, there is a danger that what awareness is developed happens through a process of curriculum osmosis, and is colored by the inevitably simplistic models of science and technology that underlie the curriculum activity.

A recent UK curriculum initiative, the Science and Technology in Society (SATIS) project, attempts to avoid this oversimplicity [2]. The project, sponsored by the Association for Science Education, has produced large numbers of self-contained, short teaching units. These use case-studies, simulations, projects, and games, based on real science and technology events and issues, to present students with the practical, social, economic, and ethical issues involved. These are typically produced by small teams of teachers working with practicing scientists, technologists, or industrialists expert in the relevant field. Evaluation shows that such short units can be successfully interspersed into conventional courses that would otherwise be missing a significantly explicit "awareness" element.

"Doing" Science and Technology

A dominant feature of English curriculum development over the last 30 years has been the conviction that, at least in practically related subjects, learning in the subject is best achieved through "doing" the subject. One of the most overworked slogans has been the probably mythical, and allegedly Chinese, proverb: "I hear and forget, I see and remember, I do and understand." Although there is certainly some research support for the efficacy of this general approach, most associated developments in science and technology have been predominantly based in pragmatic enthusiasm [19]. (It is interesting that only two recent major science curriculum projects in the UK have had a strong base in learning theory: the "Science 5-13" project [22], explicitly based on Piaget's work, and the "Integrated Science Project" [21], which rigorously applied Gagne's model of learning. Both have been very influential on curriculum debate but have had minimal take-up in schools.) There is no doubt that this emphasis on "learning by doing" has transformed for the better the practice of many teachers of science and technology related subjects; but these transformations may owe as much to rekindled enthusiasm and the Hawthorne effect as they do to curriculum theory and research.

In English science teaching, this practical approach to science, with its classroom model of "the child as scientist," [13] was galvanized by the Nuffield Foundation's science projects of the 1960s, and is exemplified in two aspects of national science policy. In the 1970s central government established an Assessment of Performance Unit (APU) to monitor children's performance in key subjects of the curriculum. The assessment categories used for science [4] were:

1. Using symbolic representations
2. Using apparatus and measuring instruments
3. Observation
4. Interpretation and application
    i. of given and everyday information
    ii. of taught science concepts
5. Planning investigations
6. Performing investigations.

This model of what is worth assessing in science education clearly views "practical investigation" as the culmination of school science activity. The "whole investigation" is the opportunity for the student to bring together practical skills and theoretical understanding in order to operationalize a scientific problem, select and control appropriate variables, design and execute relevant experiments, and analyze and evaluate the results. This research-focused model of science has been carried forward into the National Curriculum for science, despite government concerns that learning of scientific fact and theory may be somewhat undervalued. The National Curriculum [10] defines two major elements ("Profile Components") for science for all pupils from the ages of 5 to 16:

1. Exploration of science
2. Knowledge and understanding of science, with the former carrying between 30% and 50% of assessment weighting.

A similar progress can be traced in English technology education over the last 30 years, although the detailed history is immensely more complex than for science, with educational and political demands from the diverse traditions of the crafts, engineering, the sciences, and design, among others, for a legitimate say in the nature of technology in the curriculum [8,19]. Despite this complexity, a common theme has been the stress that the essence of the technology curriculum should be "doing technology," based on sometimes conflicting models of what it is that technologists do. As with science, the formulation of the Assessment of Performance Unit (APU) has been highly influential [3,5]. This focuses on the development of students' capabilities in three areas:

1. *Procedural*: the ability to select and apply appropriate design and technology procedures in identifying tasks, investigating them, generating and developing solutions, and evalu ating them.
2. *Conceptual*: the ability to demonstrate and exploit platforms of conceptual knowledge (*e.g.*, of materials, energy, control) in resolving the task that is set.

3. *Practical*: the ability to manifest (*e.g.*, through sketches, diagrams, notes, and models) the depth and development of thinking.

(Although the A.P.U. identified four value areas relevant to technological activity--technical, economic, aesthetic, and moral--these have not so far been pursued in assessment surveys.)

This emphasis upon the task as the heart of school technology activity has now found a statutory base in the National Curriculum for Technology [11]. This identifies two "Profile Components":

1. Design and technology capability[1]
2. Information technology capability.

The former is defined in terms of four attainment targets:

1. Identifying needs and opportunities
2. Generating a design
3. Planning and making
4. Evaluating.

The accompanying programs of study and criteria for assessment elaborate this task-based model of technology. Skills of drawing, modeling, and making are important only in so far as they contribute to successful task execution. Similarly knowledge about specific materials, energy resources, or control systems is only necessary if it resources the task. Knowledge and skills are seen as resource bases to be drawn upon as required in exploring and developing possible designs to meet a need.

Science Investigations, Technology Tasks, and "Capability"

Comparison of these two process-driven modes of curriculum activity--the scientific investigation and the technological task--is instructive. They both reflect a concern to develop capability in students, to get students involved in making decisions about what is worth doing, about how best to do it, and about what needs to be learned to do it [16,20]. In these respects, they both fit into that category of activities that Black and Harrison have called "Task-Action-Capability" which, they argue, epitomize all "technological" activity [6]--an all-embracing use of the word technology which further illustrates the etymological confusion in the English curriculum debate.

Certainly the scientific investigation and the technological task share many common features, and are perhaps best seen as defining a spectrum of activity rather than distinct

[1]The curriculum terminology surrounding technology in the UK is involuted, to say the least. The current position, broadly speaking, is that Craft, Design, Technology (CDT), a reasonably well established subject area, is a subset of Design and Technology, a multi-subject area, which is a subset of Technology, which may also include computing and Information Technology. This semantic structure owes more to educational, political, and academic power struggles than it does to epistemology, curriculum theory, or common sense.

activities. When students--and adults--really become involved in them, they are activities driven by curiosity about the natural or human-made world, and a realization that that curiosity can be satisfied by purposeful combination of practical and mental action. At the science end of the spectrum, the curiosity may be predominantly on how and why something happens and at the technology end on how to make it happen differently or better. Experience particularly with younger children and their teachers suggests that from the point of view of children's developing interest and action these categoric distinctions are largely artificial.

What paradigm of "real science" underlies this emphasis on the science investigation? Its emphasis on systematic observation, gathering of evidence, and the cautious drawing of generalizations suggests a model that is empiricist, inductive, and clarificatory and that owes more to Bacon, Hume, and Braithwaite than to Popper or Kuhn. In practice, it is probably not a bad model for giving a view of the work of the ordinary, journeyman scientist, but it does not fit that well with the more heroic model of competing theories and scientific revolutions that are often used to give spice to the teaching of scientific theory.

This curricular emphasis on the systematic science investigation can, however, make a valuable contribution to the development of good technological task activity. It can bring rigor and system to the evaluation phase of technological activity, where alternative solutions are to be fairly evaluated or where optimization of the desired solution is being attempted. Evidence from public examinations, from APU surveys, and from discussion with teachers shows that this aspect of task work does not come at all naturally to children. Evaluation of otherwise enterprising and thoughtfully executed tasks is effectively confined to what has been called "up a bit, down a bit" evaluation. One reason for this may be related to the limited experience of many teachers in these aspects. The hard fact remains that very few teachers of science or technology have been practicing scientists or technologists; and it appears that the systematic control of variables in the context of real-life situations is something that teachers find hard to translate from theory to practice.

Applying Scientific Knowledge in Technological Activity

The Problem of Transfer

This relationship between theory and practice is the final major area to be considered in this review of science and technology in the entitlement curriculum. Conventional wisdom suggests that students' technological capability will be enriched if they have knowledge and understanding of a wide range of scientific concepts that they can bring to bear on their technological tasks. Science is an essential "resource base" for technology. The wider and deeper the students' science knowledge, the more flexible and sophisticated their response to a technological challenge. If they have studied enzymes as well as catalysts, they have the opportunity to explore biotechnological as well as chemical solutions to a technological challenge. Only with an understanding of basic electric circuit theory can they be expected to use microelectronic circuits effectively. An understanding of issues of transportation surely requires a study and some understanding of basic kinematics and dynamics. So goes the conventional curriculum argument. Hence the attempts to coordinate or even integrate the school science and technology curricula. And when these attempts fail, or are not even attempted, we see the depressing features of bad science inappropriately taught in technology--hard sums in workshops--or technology badly executed in science--shoddy model bridges, inedible cakes, corrosive cosmetics in laboratories. Detailed curriculum coordination in schools between subject areas is notoriously difficult to achieve. How important in practice is such detailed content coordination between science and technology? How transferable to technology activity is scientific knowledge, and how impoverished is students' technology without it?

A number of writers have suggested that technological activity involves a significantly distinct "way of knowing" from that involved in scientific activity [15]. The latter involves "propositional" or "explicit" knowledge epitomized in academic debate, the language and mode of knowing of "homo sapiens." By contrast, technological activity involves the "transactional" or "tacit" knowledge of the doer, the inventor, the maker, "homo faber." (This distinction is caricatured in the old joke: the engineer doesn't know why his bridge stays up; the scientist knows why his fell down.) Any teacher who has undertaken technological project work with children or adolescents knows that accurate knowledge and understanding of scientific principles is rarely a prerequisite to practical invention and progress to an effective solution. The explanation of the solution may require a sophisticated treatment of those principles; the development of the solution appears often not to.

It can, for example, appear bewildering to a teacher that a student can develop, make, and refine a device that scans the bar-codes on tinned goods, decodes them, and produces a voice-synthesized output; and yet the same student can have difficulties in explaining the functions and modes of operation of the fuse and earth (ground) lead in an electrical circuit.[2] However, analysis of the procedure followed by the student in the project activity suggests that, despite its complexity, the points at which the student had to understand (in the scientific sense) an electrical or electronic concept in order to make effective progress were actually very few; and that the concepts were basic and largely qualitative (*e.g.*, the ideas of a continuous circuit, of voltage as push and current as flow, and of a voltage pulse). Much of the remaining electronic activity involved the creative interrelating of a number of known or researched functional

[2]These examples are based on the author's discussions with 14- to 16-year-old students following separate science and technology courses leading to GCSE examinations, and with their science and technology teachers.

blocks. In such an activity, the intervention of analysis of the block or the linkage of blocks can appear to be detrimental to success. One key to success seems to lie in having intelligent access to resource bases of relevant functional blocks, with sufficient scientific understanding to recognize issues of matching and linkage.

This way of using scientific principles poses major challenges for the science curriculum. It is an uncomfortable fact that, certainly for school students, "wrong" science is as good as, and often better than, "correct science" in this respect. Arisotelean physics, the caloric theory of heat, fluid theories of electricity can all deliver the technological goods (or "good-enoughs"). To deliver 20th century scientific ideas clearly, plausibly, and economically is the technological curriculum challenge to science.

An Agenda for Research

There appears to have been little systematic research into how children in practice learn and develop competence in technology, in particular how they undertake problem-solving activities and utilize their existing knowledge and skills in science. Comparable recent research into the developments of children's scientific skills and understanding, *e.g.*, the work of the CLISP project in England [14], has signaled the importance of working from the reality of children's existing conceptual frameworks, with major implications for effective pedagogy and curriculum development. As technology education as a component of the entitlement curriculum develops, and as it becomes increasingly urgent to lay a firm basis within the entitlement curriculum for the continued development of advanced technologies, the highest priority needs to be given to developing such a research base. This task should be in itself a paradigm for that blend of creative problem solving and analytical rigor which a symbiosis of science and technology can produce.

References

1. Association for Science Education (ASE). *Technology Education and Science in Schools.* Hatfield: ASE, 1988.
2. Association for Science Education (ASE). *Science and Technology in Society* Vols. 1-10. Hatfield: ASE, 1986-1989.
3. Assessment of Performance Unit (APU). *Understanding Design and Technology.* London: HMSO, 1981.
4. Assessment of Performance Unit (APU). *Science at Age 13: A Review of APU Survey Findings 1980-84.* London: HMSO, 1987.
5. Assessment of Performance Unit (APU). *Design and Technological Activity: A Framework for Assessment.* London: HMSO, 1987.

6. Black, P. and Harrison, G. *In Place of Confusion.* London: Nuffield-Chelsea Curriculum Trust, 1985.

7. Cooper, S. "Technology Across the Curiculum and its Impact on Science Teaching," *Education in Science.* 135, 10-11, 1989.

8. Cross, A. and McCormick, R. (eds.). *Technology in Schools.* Milton Keynes: The Open University, 1986.

9. Department of Education and Science. *Curriculum 11-16, Working Papers by HM Inspectorate.* London: HMSO, 1978.

10. Department of Education and Science and the Welsh Office (DESWO). *Science in the National Curriculum.* London: HMSO, 1989.

11. Department of Education and Science and the Welsh Office (DESWO). *Technology in the National Curriculum.* London: HMSO, 1990.

12. Ditchfield, C. and Stewart, D. *Technology and Science in the Curriculum.* London: Secondary Science Curriculum Review, 1987.

13. Driver, R. *The Pupil as Scientist?* Milton Keynes: The Open University. 1983.

14. Driver, R. "Restructuring the Science Curriculum: Some Implications of Studies on Learning for Curriculum Development." In: D. Layton (ed.), *Innovations in Science and Technology Education,* Vol. 11, pp. 59-83. Paris: UNESCO, 1988.

15. Fores, M.J. and Rey, L. "Technik: The Relevance of a Missing Concept." In: A. Cross and R. McCormick (eds.), *Technology in Schools.* Milton Keynes: The Open University, 1986.

16. Layton, D. (ed.). *The Alternative Road: The Rehabilitation of the Practical.* Leeds: University of Leeds, 1984.

17. Layton, D. (ed.). *Innovations in Science and Technology Education,* Vol. 1 (1986), Vol. 2 (1988), Vol. 3 (1990). Paris: Unesco.

18. Levi, P. *The Periodic Table* (English transl.). London: Abacus, 1986.

19. McCulloch, G. Jenkins, E. and Layton, D. *Technological Revolution? The Politics of School Science and Technology in England and Wales since 1945.* Lewes: The Falmer Press, 1985.

20. Royal Society for Arts (RSA). "Manifesto of Education for Capability." *RSA Newsletter,* Spring, 1984.

21. Schools Council Integrated Science Project (SCISP). *Patterns: Teacher's Handbook.* London: Longmans, 1973.

22. Schools Council 5-13 Science Project. *With Objectives in Mind.* London: Macdonalds, 1974.

23. Schools Examination and Assessment Council (SEAC). *General Criteria for GCSE.* London: HMSO, 1990.

The Role of Technology Education as an Integrating Discipline

Marc J. de Vries

Pedagogical Technological College, The Netherlands

Abstract: Science and technology in a long historical process became more and more interdependent. The nature of technology was changed by this relationship. The science-technology treaty seems to be absent in the school subject technology. There are several ways of helping pupils to understand the science-technology relationship, *e.g.*, by paying explicit attention to it in technology education. This is essential for teaching the nature of advanced technology. Relationships with non-science school subjects deserve attention as well.

Key words: materials technology, science, technological process, technology-science relationship, phlilosophy of technology.

Introduction

Modern technology and science are closely related. Some popular definitions of technology describe technology as "applied science." Later we will question the correctness of this way of defining technology, but it shows that technology today uses results of scientific research intensively.

In education, however, this relationship between technology and science in many cases seems to be absent. In most countries, technology education grew out of some type of craft education and was and still is taught by teachers who usually have no background in science. As advanced technology is closely related to science, introducing advanced technologies into technology education, the subject of this NATO workshop, will not be possible without putting into educational practice this relationship. Technology should be presented in its integrating character. And in order to make that possible, a clear view on the relationship between science and technology is necessary. In this contribution I will first make evident that the relationship between technology and science is important for defining what advanced technology is; next I will describe this relationship in historical and philosophical perspective; finally I will draw lines to education and give suggestions about how to realize the technology-science relationship in technology education. Here technology education can serve as an integrating discipline with many subjects, science in particular.

NATO ASI Series, Vol. F 78
Integrating Advanced Technology into Technology Education
Edited by M. Hacker et al.

The Role of Science in Advanced Technology

How can we define "advanced" technology when discussing the integration of advanced technology into technology education? One answer could be: as "10 October 1990's technology." In other words, the technology as we find it at the day of this presentation. In a way that is correct: advanced technology is up-to-date technology. But for education this would be a rather poor answer. It would imply that what we taught our pupils last year is outdated now. We have not taught then advanced technology, even though it may have been advanced at the time we taught it. Furthermore, the technology we teach today will be outdated tomorrow! Are we not doing a bad job this way?

Let us look for another answer. How about advanced technology as a collection of certain specific issues on which we more or less agree that we can see them as advanced? Some examples: computer technology, biotechnology, new materials technology (more examples can be found in Markert [13]). This second possible answer no doubt also contains a degree of truth. These are fields in which many of today's technological developments take place. But at the same time, some doubt arises here too. Are these fields really new? Counting and computing by technological means are about as old as humankind. Biotechnology was there at the very start of technology's history. And it is amazing to see how "advanced" the knowledge of ancient people about materials was (see Figure 1).

Figure 1. Ancient material technology -- Egyptian metal workers

A more fundamental answer to the question regarding what advanced technology is seems to be: technology as it has become more and more related to science. This relationship, as we will show in the next part of this contribution, made a thorough change in the nature of technology. And although this relationship has grown in the course of some centuries, its results still determine the "advanced" of today's technology to a large extent. Teaching this relationship helps pupils understand the true nature of advanced technology. This type of technology education will remain valuable in the future, because this relationship will remain, even though the specific issues in advanced technology will change.

The Technology-Science Relationship in Historical and Philosophical Perspective

All early forms of technology were completely based on knowledge that was gained by experience (empirical knowledge). This knowledge was not collected systematically, nor was it brought together in an abstract, idealized, theoretical framework. It was passed on from one generation to the next and, in this process of tradition, grew constantly. An example of this is the procedure for processing materials. By trying out various mixtures of materials certain material properties were gained. This way of developing technology without use of any abstract scientific knowledge lasted for many centuries [9]. The steam engine can still be regarded as developed entirely on the basis of empirical knowledge; later scientific knowledge of the thermodynamical processes on which it is based was formulated. Even today much materials technology is based on non-scientific knowledge [6].

Contacts between formal scientific theories and technological practice arise when technicians want to know the explanation of the phenomena they use in order to be able to optimize the use of these phenomena. An example of this is the making of lenses for glasses. Lens makers wanted to know exactly how light progresses from the air into the glass, through the glass, and out of it again. Their problem was that they had to cope with a lot of lens aberrations that worsened the image the lens produced. This question resulted in scientific research into geometrical optics. The results of this research were then applied to improve lens production. This process of eliminating the ancient demarcation between science and technology took place in the period of the "classics," like Galileo and Newton [18]. Before that time no relationship between the development of theoretical knowledge in "natural philosophy" and the practical knowledge gained in making technical devices was seen by, for example, the ancient Greeks.

Another example is the use of electricity in technology. Electrical phenomena had been studied long before they were applied in technology. These phenomena were just used for entertaining people. Later people realized that these forces could well be used for heating, motion, and magnetic applications, for example.

Today the relationship between science and technology is very strong [1]. Scientists play an important role in technological developments. And the other way round, technologists play a vital role in modern scientific research.

What is the nature of this relationship? In the first place scientific knowledge is used in technology and technological products are used in science. In a conceptual model of technology we can see how this is done (see figure 2) [20].

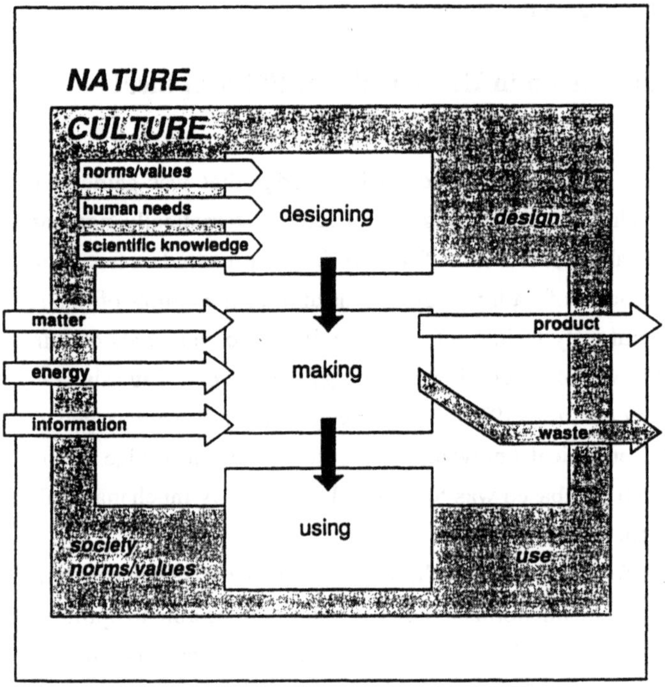

Figure 2.
Computer controlled traffic lights

The model describes technology as a process of designing, making, and using products. This process is initiated by human needs and guided by human norms and values. Scientific and technological knowledge yields both resources and constraints for this process. The process is influenced by society's norms and values, which partly determine how technology is used. The making process can be described as the transforming of matter, energy, and information into a product and waste. In the model we can recognize some moments where scientific knowledge is used:

- Designers can make use of knowledge of natural phenomena that can be applied in possible solutions to the design problem.
- In the making process (and the design of that) material properties have to be taken into account; scientific knowledge about the behavior of these materials can help here.

- When using the product natural phenomena occur; scientific knowledge can yield information on the effects of these phenomena on the (use of) the product.

But the relationship is deeper than only use of scientific knowledge in technology: science also influenced the nature of technology [19]. Let us first look at some differences in accents between science and (traditional) technology:

- In science abstraction takes place; scientists do not deal with the full reality, but only with an idealized form of it from which all kinds of second-order phenomena are detached or abstracted. In technology one cannot do that; technology has to work with the whole reality with all its second- and third- and whatever-order phenomena, because they all affect the making of a product.
- In science, analyzing is the ultimate activity: breaking the complex natural phenomena into parts that can be studied separately. In technology, synthesizing is the ultimate purpose: materials must be put together in a certain way so that a working device comes into existence.
- Finally, scientific knowledge is always a generalization. Starting with one phenomenon, one tries to develop a theory that explains more and more phenomena. It is also generally applicable. In the Netherlands the scientific laws are the same as in all other NATO and non-NATO countries. Technology is different: it deals with specification. General knowledge must be applied to one particular problem; the making of one single object. Now what happened when science started to change technology's nature? Technology took over some of the characteristics of science and became "advanced" technology.
- In modern technology problems are often solved by abstracting and idealizing first. In a model second- and third-order effects can be minimized and the behavior of the object can be studied in a more "clean" form.
- Technological problems are also solved by analysis: the problem is broken up into sub-problems and each of these is solved separately.
- More and more technological activities are aimed at gaining more general knowledge that can be applied not to one specific technological problem but to a wide range of technological problems. Even the word "technology," which literally means something like "systematic knowledge of techniques," indicates this search for more general knowledge in a discipline that originally was only concerned with practical activities. In the English language no difference is made between the practical activity and the theoretical knowledge about it; the word and concept "technik" of other languages seems to be missing here [3].

In total the process of solving technological problems in advanced technology looks like this: abstracting relevant aspects of the problem and analyzing it (breaking it into parts); then solving the abstract sub-problems, which leads to general results; next these need to be specified and synthesizes (bringing the parts together again); and finally concretizing the results.

That this process is different from the traditional way of solving problems can be seen from its results. The important role of abstraction, analysis, and generalization can be recognized from the products. Many of them are mass produced, which leads to a great uniformity (result of abstraction and generalization). Mass production is based on division of labor, splitting up the making process in parts (result of analysis). Amsterdam offers an example in the individuality of the old buildings along the canals which can be compared to the enormous uniformity found in newer parts of Amsterdam with their huge blocks of flats. The different technological processes are reflected in these products.

The results of this treaty between science and technology can be appreciated both positively and negatively [15]. Undoubtedly, the treaty has given technology an enormous influence. Technology would probably never have become the great force in our society without the changes in its nature that had been invoked by taking over certain features of science. On the other hand, the uniformity and labor division are often experienced as poor; a lot of creativity seems to have disappeared and people feel frustrated by the fact that they have lost most influence on the production process as a whole. A traditional cabinet-maker both designed and produced his products; a worker in a modern furniture factory either only designs part of the product or makes part of it. There is no feeling like "this is my piece of work." And more and more we realize the value of pieces of work of which there is only one or at least a very limited quantity But however we value the treaty between science and technology, its existence is a fact and we will have to deal with it, both in society and in education.

Yet, although technology has been influenced largely by science, it is still a discipline on its own [4]. Technology is not just applied science. There is still a long way between scientific laws and their technological application through a process of non-scientific empirical knowledge and creativity, innovativeness, and practical skills. Technology as an integrating discipline brings together the various scientific and non-scientific knowledge and skills. It is a task of design methodology to find out how the various types of scientific and nonscientific knowledge interact in the process of technological designing. At the Eindhoven University of Technology I am involved in research in this area.

In the next part I will reflect on the way we can make clear the relationships between technology and other subjects, science in particular, to our pupils in technology education.

Integrating Science and Other Subjects in Technology Education

As I stated in the introduction, in technology education the relationship with science, so important in technology, seems to absent in many cases. Perhaps this problem is less urgent in primary than in secondary education. In primary education many subjects that are separated in secondary education are still taught in a more integrative way. One could wonder of course whether pupils perhaps have a "natural" understanding of this relationship, so that lessons on it would not be necessary. But, alas, research has shown that this is not the case. On the contrary, the relationship between science and technology seems to be almost absent in the pupils' concept of technology. Research into pupils' ideas about technology has been done in about 20 countries worldwide now in the Pupils' Attitude Towards Technology (PATT) project [10,16]. No PATT study ever suggested a clear concept of the technology-science relationship in the pupils' minds. Therefore it is really necessary to pay attention to it in technology education lessons.

There are several possibilities for integrating technology with science and other subjects. Some of these are [12]:

- Teaching technology in science lessons (see, for example, Holman [8] and Martinand [14], who elaborated the teaching of material properties and technological treatments on materials in science education). In a number of countries one can distinguish a movement in science education to pay more attention to technological applications of scientific knowledge. The result of this in many cases is that a technological device is taken as a starting point for teaching science or the device is taken as an illustration of the application of scientific knowledge afterwards. This does not show the creative process of development of the device that is so characteristic of technology.
- Integrating science and technology into one subject, as happened in so-called Science, Technology and Society (STS) programs [17]. As this subject in most cases grew out of science education the objection we stated against the first option often is valid here too. Besides that, integrating the subjects seems to blur the distinctive characters of science and technology.
- Using technology as an integrative principle in cross-curricular projects [21]. A technological problem is taken as a starting point for lessons to which a variety of subjects contribute, each in its own way. Here the danger is that the process of technology as a whole is difficult to distinguish because of the many different contributions. Besides that, it is not a structural but an incidental way of integrating.
- Teaching technology as a separate subject, but in that subject paying explicit attention to the interaction with other subjects or disciplines, and with science in particular. This options seems to do justice to the specific features of technology as a discipline and also makes it possible to show this very important relationship between technology and science.

But how can this be realized? It needs initiatives on various levels:

- New curricula must be developed in which the relationship between technology and science as we have described it, is made clear to pupils.
- Teachers need to be taught and inservice trained in such a way that they have both technological skills and knowledge and a certain basis of scientific insight.
- Technology teachers and science teachers in one school should have frequent meetings to

connect their lessons so that there is a smooth link between pupils' science and technology activities.

- International cooperation can be valuable for exchange of ideas and materials in this respect. Here I would like to mention the OECD project, "Science, Mathematics and Technology Education," that was started last year. In this project member countries will deliver descriptions of successful innovative projects or developments in science, mathematics, and technology education, with special attention to the relationship between these subjects. In each case study seven themes will be dealt with: the context of the project; its processes; goals and content; the pupils' perspective; teacher training; methods and settings; and assessment [11].

If the effect of these initiatives would be that technology education became more closely related to science education, this would help pupils understand the nature of the technology-science relationship. It would also have the side effect of enhancing concept learning in technology education. Learning concepts has become an important issue in science education research recently [5]. A growing number of studies into pupils' preconceptions and the process of learning scientific concepts has been published in the past ten years. Concept learning has always been a rather weak point in technology education. Blandow [2] is one of the few people who worked on this explicitly. Hacker and Barden [7] based their technology education textbook on the concept of "system." But almost no research has been done into the way pupils could learn technological concepts like "systems" and "transmissions." The methodology that has been developed in science education research could possibly be applied for technology too. This would no doubt yield valuable information about how to improve concept learning in technology education.

Final Remark

Teaching advanced technology should never mean teaching "trendy" technology. Of course advanced technology means "10 October 1990's technology" and it deals with some fields like biotechnology and new materials, fields which we all readily agree show rapid developments. Pupils should however learn things that do not become outdated, but that remain valuable tomorrow. Giving them insight into the relationship between science and technology, and in the integrative character of technology in general, will help to educate pupils for the future!

References

1. American Association for the Advancement of Science. *Science for All Americans: A Project 2061 Report on Literacy Goals in Science, Mathematics and Technology*. Washington DC: AAAS, 1989.
2. Blandow, D. *Katalog Ausgewaehlter Begriffe der Fachwissenschaftlichen Ausbildung*. Erfurt: Paedagogische Hochschule ErfurUMuehlhausen, 1989.

3. Cross, A. and McCormick, R. (eds.). *Technology in Schools.* Milton Keynes/Philadelphia: Open University Press, 1986.

4. DeVore, P.W. *Technology: An Introduction.* Worcester, MA: Davis Publications, 1980.

5. Driver, R., Guesne, E., and Tiberghien, A. (eds.). *Children's Ideas in Science.* Milton Keynes/Philadelphia: Open University Press, 1985.

6. Gordon, J.E. *The New Science of Strong Materials or Why You Don't Fall Through the Floor.* London: Penguin, 1976.

7. Hacker, M. and Barden, R. *Living with Technology.* Albany, NY: Delmar Publishers, 1988.

8. Holman, J. *Science and Technology in Society: General Guide for Teachers.* Herts: Association for Science Education, 1986.

9. Klemm, F. *Zur Kulturgeschichte der Technik.* Darmstadt: Wissenschaftliche Buchgesellschaft, 1982.

10. Klerk Wolters, F. de. *The Attitude of Pupils Towards Technology.* De Lier: ABC Publishers, 1989.

11. Krabbendam, H. and Vries, M.J. de. *Goals and Methods in Science, Mathematics and Technology Education in the Netherlands.* Report of conference. Enschede: SLO, 1990.

12. Layton, D. "Integrality and Diversity in Science and Technology Education." In: D. Layton (ed.), *Innovations in Science and Technology Education, Vol. II.* Paris: UNESCO, 1988.

13. Markert. *Contemporary Technology: Innovations, Issues, and Perspectives.* South Holland, IL: Goodheart-Willcox, 1989.

14. Martinand, J.L. *Connaitre et Transformer la Matiere.* Exploration: Serie: recherches en sciences de l' education. Bern-Frankfurt a/M-New York: Peter Lang, 1986.

15. Monsma, S.V. (ed.). *Responsible Technology.* Grand Rapids: Eerdmans, 1986.

16. Raat, J.H., Coenen-van den Bergh, R., Klerk Wolters, F. de, and Vries, M.J. de (eds.). *Basic Principles of School Technology.* Eindhoven: University of Technology, 1988.

17. Roy, R. "The Relationship of Technology to Science and the Teaching of Technology," *Journal of Technology Education* 2,5-18, 1990.

18. Sarlemijn, A. "Science and Technology--Present Relations in Historical Perspective." In: A. Sarlemijn and P. Kroes (eds.), *Between Science and Technology: Proceedings of the International Conference Between Science and Technology.* North Holland Delta Series. Amsterdam-Oxford-New York-Tokyo: North Holland, 1990.

19. Stork, H. *Einfuerung in die Philosophie der Technik.* Darmstadt: Wissenschaftliche Buchgesellschaft, 1977.

20. Vries, M.J. de. *Technology in Perspective.* Eindhoven: OPEN TECH/PTH, 1990.

21. Zuga, K.F. "Interdisciplinary Approach." In: W.H. Kemp and A.E. Schwaller (eds.), *Instructional Strategies or Technology Education,* 37th Yearbook Council on Technology Teacher Education. Mission Hills, CA: Glencoe Publishing Company, 1988.

IV. Implementation Strategies

Once the need for integrating advanced technology into technology education has been recognized, the next question is how to accomplish that. What strategies are available for realizing this integration?

Mr. Anthony Gordon (Staffordshire LEA, UK) used a number of examples from educational practice in Staffordshire to make clear that problems from the real world should be selected to show the role of advanced technologies. Problem-solving models can be useful to realize the various steps that need to be taken, but should be used loosely rather than rigidly. He also showed the importance of a supportive center for teachers.

Dr. Thomas Liao (Stony Brook University, USA) in his presentation shared a number of experiences gained with projects in which he was involved. In one of these projects a list of skills that students should acquire in education was developed. The integration of advanced technologies into technology education addresses a number of these skills. Concept learning was an essential component of a second project he mentioned. Here too, advanced technologies can be meaningful.

Various approaches for learning about technology and about advanced technology as a part of that were put forward by the discussion groups. Learning about technology should start at an early age. The program should be gender sensitive, so that both boys' and girls' needs and interests are addressed. Social and ethical aspects should not be overlooked. Possibilities for learning about advanced technology in a low-cost manner should be explored. The content of technology needs to be updated regularly. An integration of technological and other kinds of knowledge should be aimed at. Changes in the school curriculum can only be realized when appropriate changes in the education of teachers are made.

Implementation of Technology Education in a Large School System in England

Anthony T. Gordon

Staffordshire Education Authority, UK

Abstract: The growth of interest in technology education over the last 20 years has failed to address the issue of implementation methods. This paper summarizes one approach that embraces curriculum identification, curriculum development, teacher training, and school-based support (including resource provision) in the context of a mandated technology education curriculum.

Key words: teacher training, curriculum development, resources for technology education, national curriculum, curriculum

Introduction

The implementation of a large school system approach to an educational initiative requires answers to a series of preparatory questions:

1. What is the national framework for education? What local curricular requirements exist from the school system? Are there other national,local, or community related issues to do with the relationship between education and the world of application?
2. What mechanisms exist to bring about change in the teaching methodology employed?
3. What resources are available to support change, or funding for new resources?

The responses to such questions will vary from country to country, and from system to system within each country. In the context of this paper, it is necessary, therefore, to give some indication of the situation as it is applied to the school system being considered, namely, Staffordshire LEA, England.

The Curricular Situation

The four countries making up the United Kingdom have independent approaches to the education of school students, K-14. For all intents and purposes, England and Wales can be considered as one; Scotland and Northern Ireland have largely autonomous educational

NATO ASI Series, Vol. F 78
Integrating Advanced Technology into Technology Education
Edited by M. Hacker et al.

policies and systems. This paper focuses on the changes brought about at a local level as a result of the implementation of a National Curriculum in England and Wales.

In 1988, the Education Reform Act introduced a required curriculum for students K-10 in a number of key curricular areas (English, mathematics, science, design and technology, history, geography, modern foreign language, art, music, and physical education). The governmental process resulted in the provision of a legal framework for the curriculum, but without any clear indication of the content or outcomes or of the assessment procedures to be used in teaching and monitoring of national curriculum subjects.

This omission led the government to charge the Department of Education and Science (DES), the department responsible for the oversight of the national curriculum implementation, to provide some degree of detail to ensure that the desired outcomes of learning might be achieved. It instituted in 1988 a major program (which is still continuing in 1991) of detailing the content that might usefully be taught during the four key phases of education as described in England and Wales: early elementary UK R-2, US K-1; junior: UK 3-6, US 2-5; lower secondary: UK 7-9, US 6-8; upper secondary: UK 10-11, US 9-10. In addition, the student performance levels were detailed in harmony with the taught experiences and structured to allow judgment of student performance on a continuous scale, 1-10, for each of the curriculum areas required. This characteristic of education in Europe needs to be stressed for the benefit of US readers: the notion of developing a curriculum and at the same time developing the associated curriculum measures is almost universal.

The task of detailing the curriculum content and the associated assessment levels was left to the subject working parties. Groups of acknowledged curricular experts from all phases of education, community, and business and industry were brought together to produce the curricular requirements for each subject. The resulting documents provided the general guidance needed by teachers, schools, and school systems wishing to implement the curriculum. The level of detail was not sufficient, however, to allow the documents to be used for generating a detailed curriculum: the average length of each subject report was only 125 pages, within which the curriculum, the content, and the measures of performance, K-10, was described. The various subject groups each published an interim and final report. After consultation, these led to the legally defined curricular requirements for all students.

The assessment mechanisms in use in England and Wales were also overhauled in the same legislation. The secret garden of the English school curriculum, whereby the activities of the teacher were not monitored to determine any aspect of student learning was brought to a close. The government established a consistent approach to assessment in each of the national curriculum subjects and a framework for national reporting on student progress at four points: Years 2, 6, 9, and 11 (US grades 1, 5, 8, and 10).

This move required that the country adopt a view of educational assessment that is more common in Europe, namely, the milepost or continuous recording system model. In such a program, the work of the student in every curriculum area detailed as a required component by the governing agency (in our case, the government) is metered at a number of key points by the use of national assessment examinations. Each examination is created to match the desired curriculum and is in harmony with the learning outcomes for the subject programs. This approach is based upon the premise that the curriculum should drive the testing mechanisms and that satisfactory measures are available to determine the suitability of the outcomes. In the UK, the range of outcome measures varied according to need, employing approaches such as essay work, project activities, and portfolio recording methods.

This use of "realistic measures" in Europe and the rest of the Western world is in stark contrast with the US (and with some provinces in Canada which have adopted US approaches) where the testing program is not related to the curriculum requirements and relies heavily on multiple-choice questions.

The assessment methods adopted were somewhat unusual in that they fostered a range of means to assess student performance. Up to this point the UK had almost locked into the "essay response" as the sole means of determining student performance. The US, with its almost total commitment to curricularly unrelated, mechanized, multiple-choice-based testing, might envy our dilemma: a curricular-related performance measure system, but with only one approach to outcome measurement. With considerable support from the Department of Education and Science, the working groups avoided this trap and inculcated a remarkably fresh and vital range of measures and associated outcome statements, all of which were couched in terms of student performance rather that objective-related statements (often the atomistic components of student learning that fuel the multiple-choice testing machine.

At the same time, the government gave considerable support to the use of assessment methods more in tune with the needs of students, such as portfolio, project, and essay responses. The adoption of "realistic" assessment methods in all the national curriculum areas followed naturally from such a steer.

Curricular Implementation

It is one thing to produce a mandate, an entirely different matter to ensure that the mandate is followed. In the case of England and Wales, the charge fell, by implication, to the education authorities. As the employers of the teachers and the administrators of the schools, the local education authorities are the natural focus for implementation. The government, while not wishing to fund implementation, requires an agency that will oversee the process; the teachers also seek a supportive structure to enable them to undertake their duties.

Faced with such a mandate, it is interesting to see how an employer of some 9,000 teachers, K-12, undertook the charge of fulfilling the needs of the teachers according to guidelines established by the government for one curriculum area.

Staffordshire is a large rural county in central England, one of some 104 such authorities discharging all civic duties including police, fire, social services, housing, highways, consumer services, libraries, and, last but not least, the education service. The education department serves the needs of about 160,000 students, located in some 450 primary schools (ages 5 to 11), 100 secondary schools (ages 11 to 18), and 9 vocational colleges (age 16+).

Staffordshire, with an already established program of support for technology education, was well placed to provide the framework for such implementation. Prior to examining the issues in question, it is worth examining the background to the program.

Background to the Staffordshire Program

The technology education program, which began in 1988, is aimed at the establishment of a technological component in each child's curriculum throughout compulsory schooling from ages 5 to 16. It was undertaken by the introduction of a phased program (over seven years) of curriculum development, staff training, and resource provision for all schools, all pupils, and many teachers.

The program, with a budget of $12 million, is implemented by a central team of a director, 15 support teachers, and administrative and technical support staff. The team is based in a Technology Centre built for this purpose and equipped at a cost of $250,000. Two mobile resource units (converted buses) are also used to provide local support for general and advanced developments.

The technology program formed the basis of the implementation by examination of the following issues:

(i) the implications of the curricular requirements
(ii) the implications of the assessment requirements
(iii) the teacher preparation required to support (i) & (ii)
(iv) the resources required to deliver the curriculum.

The first of these, the curriculum implications, was dealt with by the establishment of teacher development groups, arranged by phase and interest (*i.e.*, by elementary and secondary, and by subject for the secondary phase). These voluntary groups comprised active, interested teachers noted for involvement in professional activities, focused by the county technology education team. The charge faced by each group was to align current good practice with the National Curriculum mandate. The resulting outcomes from each group formed the basis of the initial curriculum support program. This phase, although vital in terms of bringing together the most innovative of current activity with a national mandate, continues to function as a means of reviewing the curricular outcomes in schools and adapting to variation in local, county, and governmental requirements.

The curricular outcomes of the groups provided a degree of specificity that enabled teachers to consider the implementation of the National Curriculum. It is tempting to think that the process was complete at this stage, in that teachers in schools could have been provided with curricular guides for technology education. To have done so, however, would have ensured that the teachers had little opportunity to develop any sense of ownership of the curricular materials. The fostering of a locally implemented curriculum was and still is foremost in the minds of the team, and so another phase of local dissemination and development was instituted to ensure that schools were involved in the debate about the nature, form, and delivery of the curriculum.

In the elementary schools, the whole school staff became involved in the debate and the development of the technological curriculum. Groups of nearby elementary schools and the local high school were encouraged to collaborate to foster community based developments.

In the high schools, although it was not possible to involve every teacher in the associated curriculum development and the debate, a broad range of experience and subject areas were included so that every high school arrived at a negotiated curriculum, in conjunction with support from the technology education team, which contains many strands of technological experience.

At the conclusion of the negotiation phase, each school, in conjunction with neighboring schools, worked to produce an implementation plan identifying the nature of the technological curriculum, showing the curriculum organization and delivery methods, and specifying student competencies at ages 7, 11, 14, and 16. The plan formed the basis for the production of detailed teaching programs, as well as a guide for the technology education team in providing external support activity such as staff training, resource allocation, and curriculum development assistance.

This phase, giving continued support to the evolution of a locally developed technology education curriculum, is still continuing.

Training

The identification of the curriculum and the local adoption of same imply that the training necessary to prepare teachers to deliver the curriculum will be given. The technology education program is responsible for the training. Teacher training is undertaken at the Technology Centre and is based upon a negotiated program of need identification.

Every school involved in negotiating and implementing a technology program has to identify staff training needs. The needs are scheduled in short, medium, and long term. The initial short-term training is largely undertaken in the Centre. The follow-up to this training is school based and is intended to support the teacher "in the classroom." This would appear an expensive use of the specialist trainers, a limited resource, but evidence indicates that school-based training is often more appropriate to needs of the institution and its teachers and, consequently, more effective than repeated, conventional external courses.

The medium-term needs of the schools, usually associated with raising levels of expertise, would tend to be dealt with by attendance at county and national courses. The long-term needs, usually those associated with planning and sustaining a major initiative, are dealt with on a school-by-school basis.

Although training continues to be the key to successful implementation of the program, the in-school coordination of the technology education activities and the associated training is vital for the concept of a flourishing school-based implementation.

This concept has led to the establishment of school coordinators for technology education in every school. In almost all cases the coordinators lead the development of the curriculum and counsel staff on identification of training needs. In addition the coordinators serve as liaisons with the technology education program team support staff.

Resources

Once the curriculum plans and the teaching programs are agreed to by the Authority, the provision of facilities and resources for technology education can commence. The Authority purchases and distributes to every school the "core" resources to establish a viable technological curriculum. The exact specification of the final package, however, will depend upon the individual school's interests and expertise. Existing provision is also taken into account.

In the elementary schools, the resources are assembled "resource" trolleys or carts. It is not necessary or wise to alter the elementary learning environment, so the trolleys are a flexible answer to a resource need. Being mobile, they can be used and deployed in a number of ways to suit the school. The cost of the trolley plus equipment is about $2,000 per school.

The high schools are to have a Technology Studio in which mechanical, electrical, pneumatic, and gas services and design and manufacturing capability are installed. The cost of a fairly well-equipped Studio is typically $50,000. The key to the use of the Studios is the flexibility and the ease with which many subject areas can gain access to designing, investigating, making, and testing facilities. The most extensive use will be, however, in supporting courses aimed at equipping pupils with technological capability.

The school staff should be involved in the debate regarding the final make-up of the resources provided for the school. From such negotiation one can foster emerging technologies such as bio-tech or chemo-tech in those institutions where expertise resides and could profitably develop.

Conclusion

With such curriculum support, training, and resources provided as part of the phased development of technology education in Staffordshire over a period of seven years, every school should be in a position to face a future in which technology education figures in a locally determined manner to a nationally developed mandate.

Pre-College Technology Education and Instructional Technology: Preparing Students for the Workplace

Thomas T. Liao

State University of New York at Stony Brook, USA

Abstract: Two technology education programs are reviewed to determine how well they are designed to prepare students for the workplace or engineering careers. A list of workplace related functional skills is used as evaluation criteria for the assessment. Another aspect of this review concerns how contextual learning has been implemented. Both programs were judged to contain instructional activities that use advanced technology effectively to prepare students for the workplace or college. These programs are also excellent models of how to help students learn concepts and skills via real-world problems and examples.

Key words: curriculum, instructional technology, engineering, functional skills, enabling skills, contextual learning

Need to Prepare Today's Students for Tomorrow's Workplace

> Nothing is more important to the future of this country than the effective education of our young people. An informed and literate nation is essential to our nation's productivity and standard of living. We can't afford anything less than a well educated and well prepared workforce [4].

William E. Brock, Chair
Secretary's Commission on Achieving Necessary Skills
U.S. Department of Labor

On April 5, 1990, US Secretary of Labor Elizabeth Dole announced that William E. Brock, a former secretary of labor, will chair a new Secretary's Commission on Achieving Necessary Skills (SCANS). The above quote was a part of a statement made by Brock on the occasion of his appointment as chair of SCANS. To highlight the need for SCANS, Dole said:
"Skills levels for today's jobs are increasing, yet 700,000 young people drop out of high school each year, and another 700,000 graduate without the skills necessary to survive in the workplace. Further, some estimate that as many as one-fourth of our current work force faces substantial literacy problems."

NATO ASI Series, Vol. F 78
Integrating Advanced Technology into Technology Education
Edited by M. Hacker et al.

The focus of SCANS is to develop guidelines that define the skills needed to get and keep entry-level jobs for high school graduates who are not planning to enter college--"the forgotten half" is the apt terminology of the William T. Grant Foundation Commission on Work, Family, and Citizenship [1]. Many American secondary school students consider school a waste of time. The complaints are familiar to parents and teachers: "It's boring." "Why do I need to learn this?"

Even students who plan to go to college have difficulty seeing the relevance of what they are studying in the classroom. For example, in studying science and mathematics they are not provided with interesting examples of how concepts from these disciplines are used in various fields of engineering. Thus, there is a need to provide secondary school students with concrete and meaningful examples of how what they are studying relates to future employment and to college programs that prepare them for professional careers.

Conceptual Approach to SCANS Project

On August 9 and 10, 1990, this author was invited to join 16 other professionals representing various industries, educational institutions, and government agencies to review a draft document that describes the conceptual approach to the SCANS Project. We met with a member of the Commission and staff from the US Labor Department who were working on this project.

The preliminary thinking in the draft document and refinements that came out of the August meeting provides valuable insights for analyzing the role that advanced educational technology should play in emerging secondary school technology and engineering education programs. This paper will present an overview of the preliminary findings of the SCANS Project. Two New York State secondary school programs that use advanced educational technology will then be reviewed to determine how well they carry out the recommendations of the SCANS Project. One of the programs is a middle school (grades 6 - 8) technology education program. The other program is a pilot one-year course, entitled "Principles of Engineering," that is designed to introduce 11th and 12th grade students to the study of major engineering concepts and how they are applied to engineering problems and case studies.

At the "meeting of experts," a working document developed by the SCANS staff was reviewed. This paper, "Conceptual Approach to the Definition of Necessary Skills in Work," contained the following opening paragraph:

> The two basic objectives of SCANS are to (a) recommend the skills that students leaving high school should have in order to perform effectively at work (and be capable of continuing to learn on the job and progress in their career), and (b) provide a reliable means of assessing those skills at different levels of performance. The skills needed for work comprise only a part of what students should learn in high school for fully participating in society as citizens, parents,. and individuals--however, a very important part. All indications are that the capabilities of the students coming out of high schools fall considerably short of the increasing demands of the workplace [5].

In order to operationalize the identification of necessary skills and the means of assessing attainment of these skills, the SCANS Project is focusing on three major aspects:

- identification and description of functional skills,
- identification and description of enabling skills, and
- learning in context.

Functional Skills and Enabling Skills

Existing research shows that the functional skills of people at work involve the exercise of different and more complex capabilities of thinking and behavior than what most students are normally taught in schools. In order to perform functional skills, a set of enabling skills is needed. These are the more familiar verbal and quantitative competencies learned in schools.

One of the problems in today's classrooms is that enabling skills besides the main part of the curriculum are often learned in a vacuum. What is needed is more emphasis on the study of generic functional skills that involve the performance of multiple enabling skills. For example, the ability to manage resources (time, finances, materials, and people) effectively can certainly involve, to varying degrees, many different acts of reading, writing, quantitative analysis, etc. The group of experts that met in August was charged with the task of identifying a list of 20 to 40 functional skills that satisfied the following criteria. The functional skills must:

- be important to the effective performance of a broad range of jobs
- adequately represent the complexities of work
- encompass a wide range of enabling skills
- have a hierarchy of performance levels
- be testable, so that a person's performance can be evaluated
- be teachable, so that instructional programs can be developed

After much brainstorming and discussion, a tentative list of 27 functional skills was identified. The 27 skills were grouped in five major skills categories as follows:

I. Resource Management: Identifying, organizing, planning, and allocating resources
  A. Time: can understand, follow, and prepare a schedule
  B. Financial: can prepare and follow a budget
  C. Material: can select and use materials and tools
  D. Human: can choose and develop cooperative working groups

II. Information Management: Acquiring and using necessary information
  A. Identifying, finding, and selecting necessary information
  B. Assimilating and integrating information from multiple sources
  C. Representing, conveying, communicating information to others effectively
  D. Converting information from one form to another

E. Preparing, interpreting, and maintaining quantitative and non-quantitative records and information including visual

III. Social Interaction

A. Participates as an effective member of a team

B. Facilitates group learning

C. Teaches others new skills

D. Serves clients/customers

E. Influences (informs, explains, persuades, convinces) an individual or group

F. Negotiates to arrive at a decision

G. Works with human diversity

H. Understands how the social/organizational system works

IV. Understanding Systems Behavior and Performance

A. Understanding how system comments interact to achieve goals

B. Identifying, anticipating, and managing consequences

C. Monitoring and correcting performance--identifying trends and anomalies

D. Linking symbolic representations to real-world phenomena

E. Integrating multiple displays

V. Human and Technology Interaction

A. Selecting and using appropriate technologies

B. Visualizing operations and program machines to perform work

C. Employing computers for input, presentation, and analysis

D. Troubleshoot and maintain technologies

E. Design systems to perform complex tasks efficiently

Learning in Context

The third aspect of the conceptual approach to SCANS is that people learn skills best when they are learned in context. Establishing the connection between skills and real-world situations is equally important to the learning of both functional and enabling skills. Ideally, students are placed in a real work situation. Problems are posed and functional skills and related enabling skills are learned to facilitate the search for solutions.

Given the list of functional skills and the recognition that contextual learning is important, the next step for the SCANS Project is to generate descriptions of how the functional skills relate to various work environments. Equally important is the development of a set of evaluation instruments for determining if students have learned the functional and enabling skills. In this paper, the tentative list of functional skills and the notion of contextual learning will be used to review two technology and engineering education programs. The approach is to determine how well advanced educational technology has been used to enhance these programs. The preliminary findings of the SCANS Project will be used as evaluation criteria.

New York State's "Introduction to Technology" Program

The middle school technology education program, "Introduction to Technology," presents a broad view of technology. Students study various types of technological systems such as highway, home heating, and computer systems. They study the engineering, science, and mathematical concepts that help them develop an understanding of the behavior of these systems. How people can control technology and the effect of technology on people are other important dimensions of this curriculum. The ten units that make up this "Introduction to Technology" program are:

(1) Getting to Know Technology
(2) What Resources are Needed for Technology
(3) How People Use Technology to Solve Problems
(4) Systems and Subsystems in Technology
(5) How Technology Affects People and the Environment
(6) Choosing Appropriate Resources for Technological Systems
(7) How Resources are Processed by Technological Systems
(8) Controlling Technological Systems
(9) Technology and Society: Now and in the Future
(10) Using Systems to Solve Problems

Technology Learning Activities and Advanced Educational Technology

This course of study is designed for children generally between 11 and 13 years of age. According to educational researchers, the majority of children in this age group learns best in environments that provide concrete "hands-on" and "minds-on" experiences. One of the unique aspects of this program is the use of Technology Learning Activities (TLA's) for helping students to become technologically literate. Both low and high technologies are used to engage students in activities that help them explore the various aspects of the behavior of technological systems.[2]

Various types of advanced instructional technology are used in the TLA's. Microcomputers are used in various ways. Micro -worlds such as "LEGO/LOGO" and "The Factory" are used to help students learn about how subsystems are combined to create systems that can be controlled by computer commands. This type of computer activity also involves student in creative problem solving and allows them to construct their own models of or solutions to problems.

Students can use simple computer-assisted drawing software as a design tool. Some classes use analytical tools such as spreadsheets, databases, and graphics software to conduct "what-if" investigations. Other classes use working replicas of satellite communications systems and other telecommunication activities.

"Introduction to Technology" and SCANS

How well does this middle school technology education program relate to the preliminary recommendations of the SCANS findings? Put differently, does "Introduction to Technology" help to prepare students for the workplace? In terms of the functional skills that relate to resource management, there seems to be an excellent fit. In fact, two of the ten units of study directly address management of materials, time, and money. Having students work in groups of various sizes and implementing cooperative learning techniques would improve the fit. Although some of the TLA's address the issue of information management and use the computer as an information processing tool, more could be done to help students to develop the functional skills for information management. Although some teachers, because of the availability of advanced educational technology, do work on these skills, techniques for developing these skills need to be integrated into the curriculum and TLA's.

The social interaction skills are much more influenced by how the class is conducted than by the content of the curriculum. In order to insure that students develop these functional skills, the TLA's should be designed so that students have to work in cooperative learning groups. The only way to judge whether students are acquiring these skills is to observe their behavior in work-related activities.

The fourth SCANS category which deals with understanding systems behavior and performance is also a significant component of the "Introduction to Technology" program. In fact, systems analysis is one of the cornerstones of the curriculum. The functional skills in this category are all part of the curriculum and these skills are developed via the TLA's. The development of these skills can be greatly enhanced by appropriate computer-based learning and other types of advanced educational technology.

The functional skills that relate to the human and technology interaction category are not directly covered in the curriculum. However, some of the TLA's do address certain functional skills. Currently, whether students in this program learn these functional skills is highly dependent on what TLA's a teacher chooses to use.

In summary, "Introduction to Technology" seems to do a good job of developing functional skills in two of the five categories. The curriculum also does well in providing contextual learning. Most of the TLA's relate to real-world situations or concrete problems. Students in these classes are seldom bored, and they know why they are developing functional and enabling skills. The primary role of advanced educational technology is its use in problem solving, modeling, and data analysis. This use of technology models what occurs in the workplace.

"Principles of Engineering" Course

Currently, the New York State Education Department is field-testing a newly developed course called "Principles of Engineering." This course, designed for college-bound 11th and 12th grade students, uses the case study approach to help students learn and apply six major conceptual areas of engineering. During the 1990-91 school year, the following four case studies are being field-tested:

(1) Auto Safety

(2) Ergonomics of Communication Technology

(3) Machine Automation

(4) Structures: Emergency Shelters

This author is part of the curriculum development team and is writer for the second case study. These case studies or mini-courses are vehicles for helping students learn and use the major concepts. In order to convey the scope of this course and the six conceptual areas, the following performance objectives are reproduced from the draft of the curriculum guide [3].

Upon completion of the set of case studies in the course, students will be competent in:

1. Modeling
    a. Use of words, pictures, and mathematics to describe a system.
    b. Manipulation of models of systems through the use of apparatus, computer simulations, and mock-ups.
2. Systems
    a. Describing a system in terms of input and output.
    b. Explaining and demonstrating how a specific system is made up of subsystems.
    c. Demonstrating how a system is controlled through feedback.
    d. Contrasting open- and closed-loop systems.
3. Optimization
    a. Explaining the consequences involved in trade-off situations.
    b. Setting criteria in real-world decision-making situations.
    c. Explaining how constraints often conflict with the ability to meet the desired outcome in decision-making situations.
    d. Developing the ability to use techniques such as applying algorithms and appropriate trial and error in making decisions.
    e. Using cost-benefit analysis and cost-effective analysis in making decisions. Cost is considered as human, societal, political, environmental, as well as economic.

4. Technology-Society Interaction
   a. Applying the system of technology assessment regarding the future impact on society of the application of specific technologies.
   b. Describing the process of considering alternative approaches to the solution of technology-society problems. These alternatives fit into the categories of:
      1). Education (behavior modification)
      2). Legislation (rules and laws)
      3). Technological fixes (applying technology to the solution of a problem)
   c. Actually getting involved in voluntary action such as lobbying, recycling, and/or developing a technological fix in a real-world situation.

5. Design
   a. Participating in the design process.
   b. Considering human and environmental factors in the design of a system or device.
   c. Applying design principles such as form and function, color, balance, unity, etc. in the design process.
   d. Selecting appropriate materials when designing.
   e. Considering the effect of production capabilities, marketing, time, and cost in designing a product.

6. Ethics
   a. Considering the legal and professional responsibilities of contractual agreements and activities.
   b. Exhibiting social responsibility by considering:
      (1) Benefit or risk to society.
      (2) Benefit or risk to individuals.
      (3) Environmental risk.
      (4) Long-term vs. short-term risk and gains in making decisions.
   c. Being aware of the moral dilemmas involved in employment.
      1). Do you blow the whistle on your employer even at the risk of losing your job?
      2). How do you weigh the benefit vs. risk in making a decision?

Case Studies and Advanced Educational Technology

In each of the case studies, advanced educational technology is used to provide students with powerful tools for studying and finding solutions to questions and problems. Students are provided with examples of how engineers use computers to design and study the behavior of systems.

In the Auto Safety mini-course, several activities use microcomputers. Students use a computer simulation called "Yellow Light" to study the timing of traffic lights at an intersection.

A microcomputer-based lab uses a motion sensor to study how a model safety car is moving prior to and during a collision. A television camera is also used to videotape the collision between the model vehicle and a wall.

Two computer-based activities are used to help students learn about the ergonomics of the TV picture. In a case study entitled "Vision and Television," students first learn about "persistence of vision" by controlling the rate that two colors flash to make a purple color. Another computer program is used to help students to determine how well they see detail at a distance. These computer activities help students to answer the following questions about the TV standards that were set 50 years ago:

(1) How many pictures per second?

(2) How many lines per TV picture?

In the "Machine Automation" mini-course, students use the Fischer-Technik modeling system to build models of industrial systems that are automated. Students build subsystems and combine them into systems for carrying out certain functions. A BASIC program is used to provide computer control of the system that has been designed and built. Besides program control, simple feedback control can also be used to automate the systems.

Students use CAD (Computer-Assisted Design) systems to help them design an emergency shelter for hikers who are trapped on a mountaintop. In this case study, students also experiment with insulation properties of different materials. Finally, a spreadsheet program can be used to carry out what-if investigation about the performance of the shelter.

"Principles of Engineering" and SCANS

Although the SCANS Project findings mainly deal with skills for students who do not plan to go to college, they can be used to assess courses for college-bound students. Many of the functional skills that are needed in the workplace are also needed for success in college. Contextual learning is the primary design criterion of the "Principles of Engineering" course. Using the case study approach makes the learning of engineering concepts relevant. Using computer technology to support instruction provides students with examples of how technology is used in the workplace.

Four of the five functional skills categories are well represented in the "Introduction to Engineering" course. Students will learn functional and related enabling skills that will serve them well in college. The only category that is not addressed explicitly is social interaction skills.

This limited assessment of two technology/engineering programs shows that they are designed to prepare students for employment after high school or for engineering careers after college. When the SCANS Project develops assessment instruments, this preliminary finding can be verified with students who are in these two programs.

References

1. Hamilton, S.F. "Is There Life After High School?," *The Harvard Education Letter*, July/August 1990.
2. New York State Education Department. *Introduction to Technology, Curriculum Guide and Technology Learning Activities.* Albany: 1987.
3. New York State Education Department. *Principles of Engineering, Curriculum Guide and Engineering Case Studies. Albany:* 1989.
4. U.S. Department of Labor. "Labor Secretary Dole Announces Workplace Skills Commission." News Release, April 5, 1990.
5. U.S. Department of Labor. "Conceptual Approach to the Definition of Necessary Skills in Work." Working Paper. Washington, DC: 1990.

V. Selecting Advanced Technologies to be Integrated

Integrating advanced technology in technology education implies the selection of certain examples of modern technologies that will be given a place in the curriculum. This selection is necessary because of the enormous number of possible issues that could be integrated. But what criteria should be used in this selection process?

Dr. Martial Vivet (Universite de Maine, France) was involved in a project in which microelectronics and robotics were used to stimulate problem-solving activities by students. In such an approach advanced technologies are integrated in activities that are characteristic for technology in general.

Dr. Gregor Tyrchan (Wuppertal University, Germany) did a study of the integration of CAD/CAM and other information technologies into technology education in various European countries. In most European countries a process of introducing these issues into technology lessons can be seen now.

The general outcome of discussions was that advanced technology issues are not an aim in themselves in technology education. These issues should be selected in such a way that teaching and learning general technological concepts -- such as the concept of systems -- and principles are stimulated, so that students gain an understanding of the nature of technology. This implies that not all advanced technologies need to be dealt with. A "window" approach, in which students get a general impression of advanced technology without necessarily looking at details, can be appropriate. Presenting all technologies would not even be possible because of the great variety and rapid developments. In addition to the examples that had already been given in the two presentations, new materials processing and biotechnology were mentioned.

Learning with Micro-Robotics Activities

Martial Vivet, Jerome Bruneau, and Christophe Parmentier

Universite du Maine, France

Abstract: The presentation concerns micro-robotics activities in classrooms. LOGO is the language used to drive the devices. From a distinction between "problem-solving situations" and "exercises," we reach a concept of "pedagogical degree of freedom." We then underline the central role of activities around a project when learning. An example of such a project (a machine for automatic packaging of fuse testers) is given to illustrate the pedagogical attitude. This attitude replaces other pedagogical possibilities that exist with micro-robots.

Key words: learning environments, micro-worlds, micro-robots, problem solving

Introduction

We describe here our proposal to train pupils in technology. The approach has been tested in classrooms in French colleges. It is based on use of micro-robots in problem-solving situations. We have shown in earlier papers [26-28] the different abilities which can be tackled and the different pedagogical attitudes for a teacher with such tools.

We try to come back here to a characterization of problem-solving situations, the concept of project. The difference between exercise-situations and problem-situations is underlined and we explain that from the learning point of view, problem-situations are more powerful if we work in open space problems. We then give a first characterization of micro-robotics as a kind of micro-world appropriate to help people learning technology.

Problem-Situations vs. Exercise-Situations When Learning

The dichotomy between problem-situations and exercise-situations comes from the conception we can have of learning itself. For most cognitive psychologists, learning is tied to problem solving. Problem-solving potential for learning is present only if relevant problem-situations are available. So it is interesting to build learning environments from this concept of problem-situation. The hardware and software used to do so have to be chosen carefully to approach learning in technology.

NATO ASI Series, Vol. F 78
Integrating Advanced Technology into Technology Education
Edited by M. Hacker et al.

In a classroom, very often there is only an univocal contract between the learner and the teacher. The goal for the learner is to reach a good solution to the proposed exercise, this exercise being a direct application of what has been taught by the teacher. The learner tries to reach the solution the teacher is seeking. The teacher is only looking for feedback from what s/he previously said. More or less the teacher is the author of the solution. The logic behind this method is that of transmission of knowledge. We have in such a case a very low pedagogical degree of freedom for the learner. Basic choices are in the definition of what is transmitted, this being relevant for the pupil or not.

This approach seems to us not appropriate for a global approach to a technological training. The problem is to motivate and show the necessity for some pieces of knowledge to become effective in a given technical world. We need to instill technological creativity and teach students to make decisions in technical environments and to internalize concepts by reconstructing some of them.

As told by J. Piaget, "comprendre, c'est inventer. On reconstruit par reinvention. Et il faudra bien se plier a de telles necessites si l'on veut dans l'avenir faconner des individus capables de production ou de creation et non seulement de repetition."

The analysis of the relations between the knowledge corpus and school language vs. pupil language can explain causes for failure in exercises of direct application for most of the children [3]. For the children, the knowledge value is only in what it allows them to do. The language is implicit and directly linked to everyday life; it allows description of concrete situations. For the teacher, the knowledge is an object, organized in separate items, each being teachable with an explicit language carrying knowledge in itself. The applicative exercise is coherent with this logic where knowledge is first taught to the pupil with the request for him or her to apply it in a local situation.

A problem-situation is fundamentally different because it is not the direct application of transmitted knowledge. It is an engine for activity helping the pupil to acquire abilities by doing. Such competences are generally at first expressed in an implicit way by direct manipulation of objects and informations. They can be transcripted under a more formal aspect [20]. Then they can be explicated and socialized with teacher (or co-learners) mediation during synthesis phases. The pedagogical goal is here to allow for the pupils an activity leading them to express in different forms their solutions, their steps in the solving process. This has to be done before explication phases and transmission to others (these being the teacher, co-learners, or a machine).

Generally a problem-situation is not specified under a unique and precise problem form. A set of goals leading to a set of sub-problems which are clarified during an analysis phase is given. This lead to sets of solutions which can be discussed according to different criteria. For example, in technology, criteria like efficiency, maintenance, security, and economy can be used. For each sub-problem, the learner must be active to find solution(s). An experimental attitude, proceeding

sub-problem, the learner must be active to find solution(s). An experimental attitude, proceeding cautiously, with phases for assimilation and adaptation, must be respected. One solution among others which can be discussed is then produced with a language strongly linked to the situation, defined during the activity, and relevant for the available devices.

Confronting a learner with a problem-situation is relevant only if:

- the situation is meaningful for her (him)
- the situation has a pedagogical sense for the teacher (according to the topics s/he has to teach)
- one or more languages are possible to express the problem-situation and the solution(s)
- the pedagogical practice is based on experimental groping, means-ends analysis, with transfer of knowledge done only at the right moment--when needed!

Our work in micro-robotics is a search for problem-situations pertinent to helping people to learned basic concepts in technology. Validation of the approach is continuing with children, and we are now using the same approach to retrain low qualified adults, helping them to face reconversion problems in better conditions. We are engaged in such experiments with a firm facing reconversion problems to introduce computer integrated manufacturing (CIM) [29].

Micro-Robots to Build Micro-Worlds

Pedagogical Degree of Freedom

Our approach is largely based on the concept of micro-world as described by S. Papert [12]. So, we are looking for concrete problem-situations allowing a more or less richness and diversity in activities. In micro-robotics, the micro-world can be understood as a set of physical devices and a programming language (we use LOGO, for which we have written specific technical primitives).

Robot-based micro-worlds can offer various pedagogical degrees of freedom [26] in comparison to mechanical degrees of freedom characterizing industrial robots. Our practice in classrooms [2] has now clearly shown that the diversity of potential activities and the possibilities approachable by the learners are unequal. A rule can be given here as a first conclusion for this work: "the more the robot is organized as a dedicated automata for a specific kind of manipulations, the more the language to use it is technically sufficient. Moreover, in such a case, the less rich the potential activities are, the less the pedagogical degree of freedom is."

The lack of pedagogical flexibility comes from the fact that dedicated systems are in themselves solutions to predetermined problems; technical arbitrations are yet done. In using such tools, the cognitive activity of the learner is constrained by the choices previously made by the designer. The potential problem-situation possibilities are reduced by the proposed device.

Examples

The washing machine. We can find in books to train pupils in technology the description of a washing machine as an example of a complex and structured technical object. The real problem-situation concerns the need for washing a heap of different clothes, made of different materials and of different colors, for example. It does not concern the description of a pre-build device.

Starting from this situation, an analysis can lead to the concept of classes of clothes and to specification of the different functions to apply in each case (temperature, timing, etc.). So the different programs can be explained from the problem-situation. The washing machine has not been invented to produce problems in automation. It is the answer for a given problem. We have to start from the problem itself and not reduce our activity to exercise on a given pre-described machine.

The lift. Problem-situations based on a lift and GRAFCET as the command language are very often used. A user calls from x^{th} floor to reach the yth one. Leading the pupil to write a program to move the lift pushes him or her to write a chart of constraints, taking account of priority rules, security, and coherence in moves from the user's point of view. But, the choice of GRAFCET, the language devoted to this kind of application in professional environments, is a constraint twisting the solution. The expression of the program is reduced to the kind of objects and control structures it can manipulate. During the elaboration of the chart, the pupils have to know this language to focus their solution according to the limits of such a language. The goal could be to teach GRAFCET. The goal could also be to make the pupils discover first the functions they need to solve a problem (move the lift) and later to find a possible tool to implement a solution. In any case, it can be observed that the pedagogical richness of the technical environment does not expand with the technical complexity. Worse, the more complex tools, fitting completely with a given need, do not set problems. They can help people to reproduce solutions and to reuse a solution on different problems without critical attitude. This does not help them to acquire a positive, creative attitude to face a technical problem.

One Example of Problem-Situation

The goal is here to describe an experiment in which pupils had to design an automatic machine to pack objects (fuse testers) produced in the classroom.

The Project

The context here is a course of technology in a class of 11- to 12-year-olds in a college. Pupils come from a popular suburb of Le Mans. It is important to observe that a class of 14- to 15-year-olds in the same college was also involved with the project. The older pupils had to produce the protocols of communications between the different manipulators designed by the younger pupils. So the project was an object of discussion in the school as a whole.

The project had four phases:

- economical and technical analysis with the specification of a set of charges as an output
- design of a product
- manufacturing
- commercialization

This first project for the pupils is very important because it helps them acquiring basic mechanisms and fundamental representations of a technological attitude. They are involved into total project management. This gives them a large motivation to go ahead with the enthusiasm of discovering.

In such a project, activity with micro-robots found a place in the need for packing the manufactured products.

The Attitude

During the project no course has been delivered to pupils to describe methods or techniques. The project has been rich enough in problem-situations to face major themes. Such problems are not set by the teacher but elicited and expressed by the pupils during their search. The choice of the project is essential. The problems must be neither too simple nor too complex. Learning by doing is a main aspect. The basic pedagogical problem is to structure the sequence of activities to allow responses to problems specified by the pupils and to share during synthesis periods results produced for local problems.

The role of the teacher is very different from giving courses in a usual meaning [21]. S/he can help in choosing basic tools. S/he has to present the project and explain its main phases. During the activities s/he acts as a facilitator, helping in access to knowledge (encouraging communication with others, assisting with searches for relevant information in documents, books, databases). A major aspect is in the pressure the teacher has to maintain to oblige kids to formulate and express correctly their ideas or needs. Facing the need for a correct expression to communicate is a way to impose a correct formulation of the problems. And generally in such cases, either the problem is solved during this process or the need for a well-specified knowledge item is elicited. At that moment, the teacher can act as an expert, "transmitting" a given piece of knowledge, giving and explaining vocabulary for example. S/he will be heard because the need for an answer is great at such a moment!

Responsible at the pedagogical level, the teacher has to maintain coherence during the emergence of the different notions with the goals of the specified curriculum for such courses. The coordination of synthesis phases within the project and in concert with other colleagues is his or her pedagogical responsibility. For example, people teaching French may be involved when facing the need for writing documents [24].

Useful Hardware and Software

During the project, the work room is organized with three stations for 24 pupils. Each station contains a micro-computer, a Fischertechnik interface, and a set of separate Fischertechnik bricks. A fourth station is kept free and available for any number of classes to test specific design solutions besides the design proposed by a given group. A shop with available bricks is organized at the rear of the class which allows pupils to choose additional bricks. The management by the pupils of such a shop is in itself a part of the project.

LOGO with specific primitives to manage the interfaces is used as the programming language. A set of technical sheets to allow individualized help is available. Their content concerns acquisition of specific know-how (how to connect the interface to the computer, how to connect an engine to the interface, how to drive an engine with instructions, how to connect a captor, how to produce a translation or a rotation with bricks), description of technical aspects (name and arguments of programming primitives, technical aspects on the interface...). In the project, younger children had only to design a manipulator M1 for the fuses and a manipulator M2 for the boxes. The manipulators to close the boxes and to evacuate the boxes were designed by older pupils.

Characterization of Pedagogical Micro-Robotics

The goal here is to try to characterize major aspects of what we have called "pedagogical micro-robotics." This domain can be characterized by three main components: hardware aspects, software aspects, and pedagogical attitudes.

Hardware Component

1. The permanent presence of "real" objects, physical devices made with material which can be broken or bricks which can be assembled, is the more visible aspect. Pupils can touch, and manipulate these objects. They have a weight and physical dimensions coherent with the physical dimensions of the pupils themselves: the objects can be included in cubes with edges say less than one meter long. This means that they can examine globally what happens in the whole micro-world from outside of this world. This characteristic is clearly separate from what can be done by simulation with "abstract pixels" on screens.
2. The hardware is devoted to pedagogical functions. With respect to the work by didacticians, the concepts of "transfer" and "didactical contract" are always present. So a main concern is in the study of what can be transferred from the learning point of view when

using such devices and how this can be done. The elicitation of favorable conditions to allow such transfers is a part of the domain.

3. The devices are always controlled by a computer. They are non-standard peripherals from the usual point of view in computer science. They are consecrated to "learning functions" and are designed with this focus in mind.
4. Learning potentialities concerns:
   - disciplines like mechanics, physics, electronics, mathematics, computer science [5, 24]
   - approaches like experimental approach in sciences [11]
   - concepts like flexibility in production processes [1,17]

Software Component

The basic software tools used are more like tool boxes with relevant available functionalities than software in the usual sense in computer assisted learning (CAL). We do not use directly software whose function is to teach or to verify knowledge. This kind of software is not rejected completely if some is available to help as a resource, but it generally does not fit the attitude we use in projects.

Pedagogical Component

1. The pedagogical attitude concerns mainly the construction of knowledge, the elicitation and explication of pre-existant knowledge. The direct "transmission" of knowledge is done when the need for that knowledge is clearly expressed to face a problem-situation. This transmission can be done directly by the teacher and also with resources made available to the pupils (cards and hyper-media tools, for example, can fit the approach).
2. The approach is centered on the learner. The activity is learner driven (as opposed to teacher driven!). The project, or problem-situation, is the engine inciting to activity and motivating work.
3. The computer is used as a controller for devices in its environment. From the pedagogical point of view, we shift from computer assisted learning (CAL) to computer allowing learning by exploration (CALE!). It is possible to converge here with Dickson's attitude and his "exploratorium" [4].
4. The project is a central object allowing inductive approaches and learner's activity managed into a group. The necessary socialization of knowledge is included in the activity itself.

Conclusion

Our work is an approach to helping people build their own knowledge and culture about technology. The experiments done have clearly shown to us the feasibility of open pedagogical practice based on projects and problem-situations. Teaching technology without involving the learners in a total technical project realization seems now to us to be a sham. The very high level of motivation shown by the pupils is an encouragement to go on deeper. The success we are now observing with low qualified adults confirms our feeling of the problem. But it is sure that such an attitude works well only with very motivated teachers who are engaged themselves in projects. Perhaps here are the limits of the approach!

References

1. Baudry, M. "SPF: Un environnement d'apprentissage des concepts lies a la flexibilite dans la production." Premier congres francophone de robotique pedagogique, Le Mans, 30/8-1/9, 1989.
2. Bruneau, J. "Remarques autour d'une activite de robotique en classe de 6 ieme," Premier congres francophone de robotique pedagogique, Le Mans, 8/30-9/1, 1989.
3. Charlot, B. L'ecole en mutation. Payot, 1987.
4. Dickson, P. "Environment for Interactive Learning: The Computer Exploratorium," NATO-ASI, Calgary, July 16-28, 1990.
5. Enrique Ruiz-Velasco Sanchez. "Un robot pedagogique pour l'apprentissage de concepts informatiques," Premier congres francophone de robotique pedagogique, Le Mans, 30/8-1/9, 1989.
6. Giovannini, M.L. "Robotique pedagogique a l'ecole primaire: Une premiere experimentation Lego-TC-LOGO a Bologne," Premier congres francophone de robotique pedagogique, Le Mans, 8/30-9/1, 1989.
7. Gordon, A. New Technology in Secondary Schools: Problem-Solving," Milton Keynes: NATO, November 11-12, 1988.
8. Marchand, D. "Construction et programmation par des eleves d'un micro-robot." Premier congres francophone de robotique pedagogique, Le Mans, 30/8-1/9, 1989.
9. Mercier, I. "Modelisation du processus d'apprentissage des objets techniques." Premier congres francophone de robotique pedagogique, Le Mans, 8/30-9/1, 1989.
10. Nonnon P. "La robotique pedagogique," revue BUS, AQUOPS Montreal, Mai 1987, pp. 16-19.
11. Nonnon, P. and Laurencelle, L. "L'appariteur robot et la pedagogie des disciplines experimentales," *Spectre*, pp. 34-36.
12. Papert, S. *Mindstorms: Children, Computers and Powerful Ideas.* New York: Basic Books.
13. Pelchat, R. "Le robot enfouisseur de dechets nucleaires," Guide d'activites. CARES, 10450 rue Meunier, Montreal, H3L 2Z4, Quebec.
14. Rabardel, P. and Verillon, P. "Robotique pedagogique et conceptualisation du reperage tri-dimensionnel dans l'espace." Premier congres francophone de robotique pedagogique, Le Mans, 8/30-9/1, 1989.

15. Robert, F. "Un exemple d'environnement programmable destine a des debutants," Doc. interne, INRP-DP5. Informatique et enseignement, 191 rue Gabriel Pri, F92120 MONTROUGE, France.

16. Schmitt, B. and Canals, P. "Presentation d'un produit pedagogique d'initiation a la robotique," Premier congres francophone de robotique pedagogique, Le Mans, 8/30-9/1, 1989.

17. Schoefs, Y. "Exploitation pedagogique du SPF: premier bilan d'une experimentation," Premier congres francophone de robotique pedagogique, Le Mans, 8/30-9/1, 1989.

18. Tanguy, R. "Un reseau de mobiles autonomes pour l'apprentissage de la communication." Thesis. University PARIS: VI. 1987.

19. Thornton, R. *Using Technology in the Construction of Scientific Knowledge.* Milton Keynes: NATO, November 11-12, 1988.

20. Vergnaud, G. "Concepts et schemes dans une theorie operatoire de la representation," *Psychologie Francaise*, 30-3/4, pp. 245-252.

21. Vivet M. *Apprentissage autonome, sur un usage de la technologie informatique dans l'education.* Annexe du rapport SIMON, "l'education et l'Informatisation de la Societe." La documentation francaise. pp. 201-210.

22. Vivet, M. LOGO: un outil pour une formation de base a la robotique. Colloque National "LOGO et les enseignements technologiques." Le Mans: Nov. 83 compte-rendu revue ETI-LISH/CNRS, 3, pp. 5-19, 1983.

23. Vivet, M. "Pilotage de Micro-robots sous LOGO, un outil pour sensibiliser les personnels de l'industrie a la robotique" - 5e symposium canadien sur la technologie pedagogique - OTTAWA - 5/7 mai 1986. Paru dans le livre, A l'ecole des robots. Robotheque du CESTA, pp. 195-210.

24. Vivet, M. "Driving Micro-robots under LOGO: A Way to Approach Geometry." Proceedings of the second international conference for LOGO and MATHS education. Institut of education. University of London, 15/18, July 1986, pp. 216-225.

25. Vivet, M. "Examination of Two Ways for Research in Advanced Educational Technology," NATO, Milton Keynes, November 11-12, 1988.

26. Vivet, M. "Robotique pedagogique! Soit, mais pour enseigner quoi?", Premier congres francophone de robotique pedagogique, Le Mans, 30/8-1/9, 1989.

27. Vivet, M. "Micro-robots as a Source of Motivation for Geometry," Advanced research workshop "Modelling the Student Knowledge, The Case of Geometry," NATO, Grenoble, November 13/16, 1989.

28. Vivet, M. "Which Goals, with which Pedagogical Attitudes with Micro-robots in a Classroom?", NATO ARW on Advanced Educational Technology, Pavia, Italy, October 1989.

29. Vivet, M., Parmentier, C., and Bruneau, I. "La reconversion d'ouvriers specialises un defi pour la robotique pedagogiquez," Deuxieme congres francophone de robotique pedagogique, Montreal, 26-28 August, 1990.

Integration of CAD/CAM and Information Technology into Technology Education

Gregor Tyrchan

University of Wuppertal, Germany

Abstract: The paper sketches out that CAD/CAM and information technologies are to be seen as central topics for technology education and inherent to it. The synoptical outline is given by international examples of approaches concerning the aims, themes, equipment, and results.

Key words: computer in technology education, information technology, control techology, telematics, simulation, CAD/CAM

Introduction

What I do not like doing is giving a presentation of my own ideas and activities in the field of computer aided design/computer aided manufacturing (CAD/CAM) and information technology, giving a presentation of my metatechnological base and of my specific approach developed on the green table in the university and evaluated with a more or less small class. What I would like to give is a rough trial of a synoptic report concerning the situations in different geographical areas in the national and international contexts of technology education.

For this reason I wrote letters with a questionnaire to more than a dozen institutions in Germany and Europe concerned with education to answer questions about the role of CAD/CAM and information technology in their curriculum systems, especially in technology education. And I did get answers from most of them, some short ones, some detailed with brochures. By the way, I draw attention to the specific situation in Germany where the responsibility for educational matters and policies lies with the states.

My expositions will follow a short outline of technology education as my didactical frame and--drawing on it--a synoptic outline of CAD/CAM and information technology and their integration into technology education.

Technology Education

I remember my first international, or rather European, conference in 1975 in the EC-academy at Otzenhausen The presentations of the different countries seemed to be reports of another

NATO ASI Series, Vol. F 78
Integrating Advanced Technology into Technology Education
Edited by M. Hacker et al.

world, not of the same subject taught by the presenters: different didactical legitimations, different main aims, different lanes, different conditions for the subject and—not surprisingly—a kaleidoscope of topics. The early 1970s were characterized in technology education by autonomous, regionally centered approaches based on regional pedagogical traditions, and those were mainly "craft" or "handicraft" marked inclusive of their philosophies and aims.

Mainly classic topics like patchwork, wood-working, metal-working, pottery, consideration and imitation of ancient buildings, analyses of the cycle bell, and others filled the syllabi; relative modern topics like electric switch and control systems or energy supply were the exceptions. The main philosophy was to reproduce a useful product as it existed.

Then, as we now know, approaches arose independently which reflected didactical ideas, curriculum ideas, and technology features. These new approaches were problem-oriented and process-centered; they started not with a specific engineering section, not with a specific product, but with an individual or social need. These new approaches took into account the conditions and impacts of technology on the individuals and on society. I only quote the movements started by Geoffrey Harrison in Great Britain [2], the system technology approach and the related general technology approach in the FRG and GDR [8,9], and the reports in France by Legrand which were accompanied by political, economical, and education policy changes in the countries.

As representative for those modern multidimensional technology education approaches I would point to the famous "Harrison model" now in the version of 1990 (with the background of the "National Curriculum" debate) [3].

Those approaches led to national curricula in the late 1970s or in the 1980s which have many common features.

Still further on there are states in Germany and countries in Europe which partly have a different interpretation of modern society and therefore of education in it or which are struggling for modern technology education approaches. Hopefully this workshop can help to lead to an European convergence, although there are and will remain national peculiarities, of course. And this convergence becomes more important than ever in view of the consequences of the European harmonization of 1992.

What are common features in most of the European countries and in most of the German states?

- Technology is not applied science.
- Technology is more than science, but laws of science mark out the frame for technology.
- Technology has contents and methods of its own.
- Technology has impacts on society and the individual.
- Technology education is a matter for general education.
- Technology education is a subject for boys and girls.
- Technology education is not a reduced issue of engineering.

- Technology education has its specific criteria for schools.
- Technology education has as main features the didactical series: design - evaluate - produce - maintain.
- Technology education is based on theory and doing.
- Technology education is problem-centered, not product-centered.
- Matters of concern for technology education are material, energy, information, and their artificial transformations by processes into artifacts.
- Technology education has to take technological impacts into account.
- Topics of technology education should be exemplary for technology; they should be of relevance for the individuals and/or for society.

These features are seen as items, as invariants of modern didactics in technology education as stated in the 1988 PATT conference [1]. The corresponding approaches are not only concepts of didacticians; they are also working curricula in schools and have sometimes been for years or decades.

Computers in Society and Technology Education

There is no doubt that modern societies are influenced in many ways by the presence and applications of computers. The structures, techniques, and processes of production and of services as well as in medicine or at home have changed rapidly and extensively. The impacts on these areas for the individuals and society are numerous and grave. Here are some examples. The qualifications for the employees and the manner of communications have changed. There are computerized modifications in problem solving by access to databases and electronic mail. New words and words with new semantics have arisen from the dissemination of computers and fascination with them. Computer aided technologies and computer influenced procedures and behavior are substantial parts of modern societies.

Consequently educational systems could not but pick up computers into their syllabi. There is no discussion about "whether or not." Across the states and countries there are slightly different emphases on the didactical use of computers and of the assignment to different subjects.

Following the answers to my questionnaire and to literature, I would summarize: (1) computers are essential challenges and issues in the different systems, and (2) there are different approaches.

Those states and countries in which technology education is reduced, for example, to technical drawing or craft, which are not on the lane of modern technology didactics, focus the educational computer applications mainly on mathematical, algorithmic problems accompanied by (isolated) word processing. In those states and countries with modern technology didactics we find a broad variety of educational, non-mathematical applications, representing and corresponding to the universal possibilities of computers. Here technology education is an important subject for educational computer applications among other subjects. Here computer applications are seen as an inherent consequence of technology education [6,10].

In some states and countries (France, Baden-Wurttemberg) technology education is the guide subject, the central subject for educational computer activities as a matter of concern for general education. For this education policy decision, there are some good reasons.

By the inclusion of microelectronics and computer themes into technology education, its basic ideas will not and must not suffer fundamental substitutions or changes. By using classical methods and tools and computerized methods and tools, the pupils can experience technology development and its impacts. The point is to find out the right mixture of classical and computerized solutions in technology education.

What are main topics for computers in technology education? I would cite:

- **Metrology**: intelligent sensors, multiple data logging (example: temperatures in model house)
- **Communication/telematics**: code/decode at the desk and at the telegraph, cash scanner, modems, etc.
- **Control technology**: without and with feedback in models and real objects
- **Data processing**: databases, etc.
- **Planning, drawing, designing, and dimensioning**: projects, charts, covers, drawings, designs, etc.
- **Computerized production**: CNC, robotics, storage
- **Visualization and simulation**: equipment, processes
- **Teaching aid**
- **Learning aid**

In literature I found a standard definition of CAD/CAM, but no standard of what is meant by the term "information technology." Here I would like to suggest making the above list of topics into two larger groups:

1. **CAD/CAM** comprising visualization and simulation; planning, drawing, designing, and dimensioning; computerized production, and
2. **Information technology** comprising metrology; data processing; communication/ telematics; computerized control technology.

CAD/CAM and Information Technology in Technology Education

Given the answers to my questionnaire and the brochures and literature about the curricula in the specific systems, I would summarize that the situation for information technology is better than for CAD/CAM. I will set out my findings.

Information Technology in Technology Education

Applications of **computer aided control technology** are among the oldest topics recommended and published in literature. The structures of these systems are clear; the problems readily definable; and the solutions can be deduced and derived from classical solutions (see Figure 1).

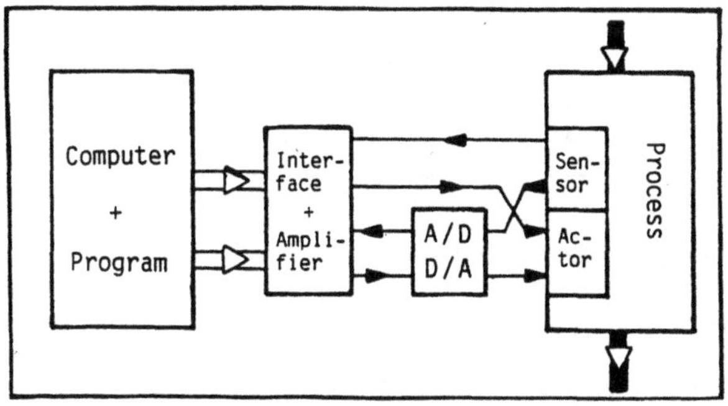

Figure 1: Control technology by computer—principal structure

In the field of computer controlled technologies there are no reasons to complain about their integration into technology education. Obviously, classical technical artifacts like traffic lights, alarm systems, heating systems, elevators, etc., can be operated easily by computers: relatively simple control unit, simple interface which could be designed and realized by students and mastered by teachers, simple elementary programming, short programs on all types of computers. The impacts of operating these systems by computers can be comprehended. The variety of proposals in literature and the syllabi is broad. There is hardly any curriculum without control technology by computers.

Computerized control technology problems can be realized by very simple interfaces which are connected to the expansion port/user port/parallel or serial port or IEC-bus including a simple amplifier. A typical exercise is--starting from the classical solution--to design the necessary interface/amplifier and realize it/them with discrete components or ICs as is shown in Figure 2 for the famous traffic lights control.

One essential part of this teaching unit is the algorithmic problem solving and the elementary programming, where rough knowledge concerning the structure of the hardware, the internal data processing, and the software/ programming language should be learned too.

In recent years, computer control technology as a field has become well elaborated by and integrated into technology education. Teachers and pupils have gained confidence in their skills, knowledge, and capabilities in operating computers with their peculiarities; they have acquired experience concerning the multidimensionality and the impacts of these computer technologies. After these successes they can go on to try other problems of computerized technologies. And indeed, that is reasonable and necessary because there are other typically technical problems, not less important, which are markedly more significant and more explosive with computers than control problems.

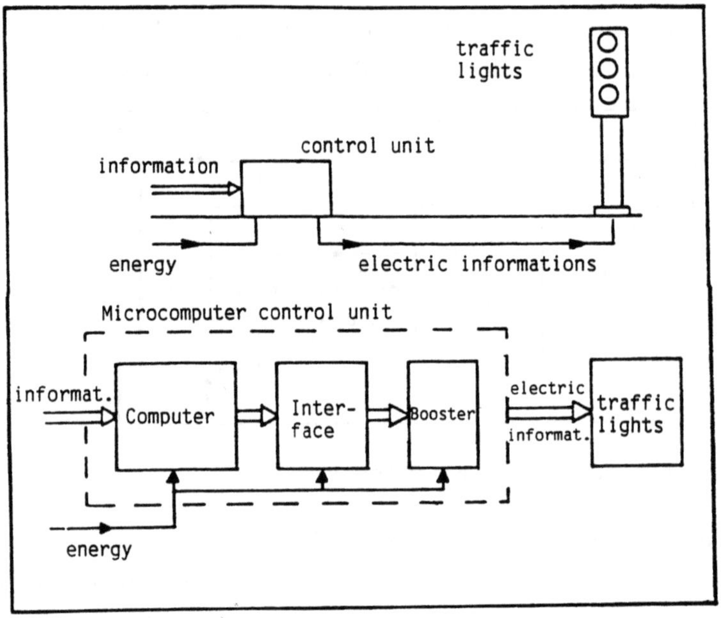

Figure 2: Computer controlled traffic

Data processing and **metrology** are marginal, very marginal in the different technology education curricula. I do not know the reasons why. Maybe the curriculum designers relate metrology more to science problems and metrology aspects to control technology. And indeed, all control problems with closed loop involve data logging, the selection of the appropriate sensor(s), the adaptation of the measurements, conversion of the signals, etc. I personally would follow this line of argument: There is no need for more metrology in technology education. Technology education--see above--must be an exemplary discipline; it is not a copy of engineering.

For me there is a similar argument for data processing. There are more appropriate subjects for this area of computer application. And we find data processing representing aspects of control problems as well as of communication/telematics.

In the area of **communication/telematics** within technology education there is a problematic situation in the countries and states. I must state that it is only in France and Great Britain that communication/telematics is integrated into technology education and has a high rank and high quality. There, the responsible institutions have for years stimulated local, regional, and nationwide activities (*e.g.*, the famous "Minitel campaign" in France, the "IT Year" proclaimed by M. Thatcher) and have thus prepared the educational systems for a future in which networks and "real-time" communication will have impacts on the individual and on society. There, central bodies have given financial and material support for hardware and software, for the development of specific software, and for its dissemination to nearly all schools, which are therefore well equipped in this section. Here we find different curricular models and projects. Electronic mail, the use of networks including central databases, and data exchanges between schools, between twinned communities, between schools and administration centers, between schools and companies (for ordering books and equipment, for example) are the "didactical everyday."

The situations in France and Great Britain are good examples of appropriate educational policy interpretation of what happens in reality in technology. The splendid situations in these two countries were favored by the rank of technology education in these educational systems, reflecting the importance of technology education for the youngsters, for the rising generations, and for adults in the years to come.

It is quite another situation for all states in Germany (East and West) and for most of the other European countries. In Germany we find only some local activities, first timid didactic trials for this important technology. Here, we have a lot of didactic activities to catch up on for the section of communication/ telematics principally and especially for technology education.

CAD/CAM in Technology Education

The situation for the area CAD/CAM within technology education is - seen as a whole - not so good as the area of information technology.

On the one hand this is surprising, because in this area we have the full palette of specific technology processes and methods with their multidimensionality and multiple perspectives, and we have them without any competition from other subjects--as happens for information technology.

On the other hand, CAD/CAM demands more knowledge and experience in the hardware and software sections, greater investments of money, equipment and time, and more in-service training for teachers and more teacher-students training, too, at polytechnics and universities.

The section **visualization and simulation** is at the moment in the testing stage. There are local activities in France, Britain, and Germany and some first publications on it, but no concepts addressed in technology curricula. Visualization is part of other topics, not an independent topic for technology education. And I think that is an appropriate role. But simulation could and should be more than part of, say, design processes on the screen as used in some CAD/CAM approaches. Simulation should exist as a topic independent of computerized design and manufacturing. There are technical problems without production aspects which can be mastered by simulation: problems of energy supply, energy transfer, problems of waste removal, consequences of faults in production lines, extraordinary stress in structures, etc. Here, technology education would have the chance to incorporate a fully new method, a fundamentally new way to teach awareness of technology and its problems. The available results of investigations in this section [5,7] could be the stimulus for further activities, for more integration into technology education.

The section "CAD/CAM" in the original meaning of the term comprises "computer aided design" processes and "computer aided manufacturing" processes. It is typically centered on design-and-making principles accompanied by technical, social, economic, and qualification impacts. The broad range of preconditions, impacts, and the spectrum of possible approaches for general education schools are well explained internationally in the proceedings of the international Wuppertal symposium about "New Technologies and School Technology" in the papers of R. Levrat, B. Harvey, and R. Oberliesen [4].

Whereas professional CAD and CAD/CAM stations require an immense amount of money, it is possible to install in schools reasonable stations on a cheaper level with a reduced range of hardware, and CAD functions, limited quality of drawings, and limited output of products. These limitations are consistent with the aims and demands of technology education didactics and with the structure of CAD/CAM.

In France, CAD/CAM is fully integrated into technology education and is seen as a main topic in the French curriculum. In well-equipped special laboratories with computers, robots, plotters, and CNC machines, including interfaces, themes and items of modern computerized engineering, and of production and workers' problems are taught. And this is taught to all boys and girls at all schools because--see above--technology education is the guiding subject for computer literacy.

Great Britain was (still is?) in a similarly good position in this section. There have been for years several LEAs with some dozen semiprofessional scholar CAD/CAM-stations. Before the National Curriculum, these courses were optional, varying from school to school. The new National Curriculum has defined technology education as a main matter of concern; and within it, design and making processes have a dominant role. Plotters, efficient computers, CNC machines, and robots have for years been standard in British schools.

The situation in Germany is not so splendid. We have didactic proposals for CAD/CAM courses in the states of North Rhine-Westfalia, Hesse, Lower Saxony, and Baden-Wurttemberg. But in these states the approaches are not implicitly integrated approaches: CAD aspects can be taught (and trained by pupils) in isolated courses where design problems are supported by computers (less than CAD); CAM problems can be taught subsequently on machine models. Mainly it is a question of financial support by the local authorities, when schools can offer integrated CAD/CAM courses or projects as they are recommended by didacticians and by curricula. Only a few lessons are scheduled for CAD/CAM items.

The products designed and manufactured in the CAD/CAM courses could be seen as mirror-images of resources available in the different countries. Whereas in France and Great

Britain the pupils can produce with professional hardware and professional software objects in plastics, aluminum, or steel for the market, the pupils in German general schools produce things like toys for themselves in styrofoam.

Let me note that we just started a common British-German project, supported by the EC as a PETRA project, to find out information and hints from British and German schools and evaluate them for an appropriate European module for CAD/CAM.

The situations in the other countries and in the other states are modest: No CAD/CAM for the general schools! It is, once again, to be seen as misjudgment by the responsible curricular institutions, as a mistake of the education policy.

Summary

The increasing numbers of international meetings and exchanges like this NATO workshop or the PATT conferences for technology education, the working cooperation of the different national asociations for technology education aiming at international contacts, and international promotion of technology education like EGTB and ITEA could and should contribute and help to pushing ahead with modern technology education didactics, with more integration of information technology and CAD/CAM into technology education.

References

1. "Basic Principles of School Technology." *Report PATT-3 Conference.* Eindhoven: 1988.
2. Harrison, G. *A Project Technology.* Loughborough: 1970.
3. Harrison, G. "Assessment of Design and Technology." In: *School Technology*, p. 12f, May 1990.
4. Kussmann, M. and Tyrchan, G. "Neue Technologien im Technikunter-richt." Report Wuppertal Symposium. Dusseldorf: 1987.
5. Steffen, H. "Computer programme fur den Technikunterricht in der S I," *Magazin fur Technik und Unterricht*, H.2/3, p. 26-37, 1988.
6. Theuerkauf, W. and Tyrchan, G. "Curriculare Konzeptionen zum Einsatz des Mikrocomputers im Technikunterricht." In: Melizinek, A. (ed.), *Medien und Technik. "Klassische" und "Neue" Medien im Technik--Unterricht*, pp. 463-468. Alsbach: 1986.
7. Theuerkauf, W. and Haupt, W. "Die Bedeutung der Computersimulation fur eine technische Bildung," *Magazin fur Technik und Unterricht*, H.2/3, pp. 15-25, 1988.
8. Traebert, W. and Spiegel, R. (eds.). *Technik als Schulfach*, Bd.1. Dusseldorf: 1976.
9. Traebert, W. (ed.). *Technik als Schulfach*, Bd.3. Dusseldorf: 1980.
10. Tyrchan, G. "Technische Bildung in Deutschland und die Mikroelektronik," *Tu* H.33, p. 8-11, 1984.

VI. Fostering International Collaboration

International contacts among those who are involved in technology education in various countries have increased significantly in the past decade. New developments are stimulated by international sharing of ideas and corporations. The integration of advanced technology into technology education is such a development. Therefore the enhancement of international collaboration was dealt with explicitly in the conference.

Dr. Kendall Starkweather (International Technology Education Association, USA) made a proposal to work towards a world federation for technology education. Such an organization would allow bringing together people and information in the field of technology education.

Another presentation on developments on a national level and the relevance of international cooperation was given by **Dr. Paul Black** (King's College, UK). In the UK the introduction of a National Curriculum, of which Design and Technology is part, is now taking place. First experiences were presented and problems yet unsolved were mentioned. Here international exchange of ideas can be useful.

During discussion the idea of a world federation was considered and various possibilities and problems were mentioned. A realistic approach was seen in establishing links between existing national associations. The possibility of setting up a database for information on developments in technology education worldwide was discussed. Several problems arise here (how to select information and how to disseminate it), but the positive aspects of such a database are evident. Finally it was stated that each network project on a national level should at least have an international component in it.

Fostering Greater Cooperation in the Development of International Technology Education

Kendall N. Starkweather

International Technology Education Association, USA

Abstract: Technology education is at a point in its history where a worldwide network needs to be created to foster communication and strengthen the field. An international agenda for technology education would assist in the coordination of efforts to advance the profession. Work on a structure for such a global federation should begin now.

Key words: global federation, teacher/student environment

Introduction

It is a pleasure to address such a distinguished group of professionals on ways to foster greater cooperation in the development of technology education at the international level. Two developments have occurred to bring us to this point in history, making this topic important for a multinational meeting. First, we now have the luxury of communication and transportation systems which allow for increased interaction and travel among colleagues. Second, professional attitudes have evolved in most countries producing a strong desire to share ideas, success, techniques, research, practices, and more. It is apparent that professionals want to obtain and share knowledge with their counterparts in other countries. Such action will hopefully result in technology education being a stronger part of the total education system in all countries.

Status

We have been fortunate over the past decade to have increased interaction among colleagues in countries around the world. My own philosophy and thought process have been influenced tremendously during this time by the writings and speeches of professionals from other nations. Fellow educators have surely experienced these same feelings in their quest to improve education. It is expected that scholars will continue this sharing process at a pace never experienced by our field. Indeed, the distance across the great bodies of water has been shortened by technological advances.

To date, nations with technology education programs have worked to advance their own agenda of educational activities. Technology educators in these countries have made tremen-

NATO ASI Series, Vol. F 78
Integrating Advanced Technology into Technology Education
Edited by M. Hacker et al.

dous advances over the years by adapting to national political issues, striving to improve their field, staying in tune with technological advances, and gaining in significance within the educational community. Technology educators around the world have held thousands of regional and national conferences within their own nations. Countless publications have evolved, providing new ideas and information sharing. Book publishers, material suppliers, and equipment manufacturers have worked in conjunction with our educators to provide educational materials.

Various nations have held joint meetings over the years that have allowed for interaction across borders. Two examples that come to mind are the North American meetings involving the United States and Canada and the meetings held by the European community. These meetings will perhaps be known as the regional meetings of the future, as interaction on the world level continues to reach a larger scope and level of significance.

As technology educators, we have now come to the point in history where a worldwide network and communications mechanism needs to be created to advance our profession. The stronger curricular areas of the future will be those that have the luxury of building on the success of the past, and, as a result, of having a world community that serves as the backbone for building the ultimate curricular area. We are far from having such an organization in technology education. However, we realize that such an organization is possible as a next step towards making a worldwide technology education community.

Currently, national agendas such as in the United States exist in structures designed to meet their proposed objectives and strategies. At best, these actions are shared at meetings, as we have experienced this week, and serve as a catalyst for those in attendance. It is time to create an international agenda for technology education that will serve as a catalyst for each nation to utilize in advancing the profession. A central task for such an agenda would be to provide the international community with a permanent mechanism for coordinating its efforts in advancing technology education.

Vision

The time has come for technology education professionals to create educational change on a global level. If we are to influence our colleagues in the other fields of education and play a dominant role in the advancement of technological literacy, we must act now. We must create a world federation which would include the members from national groups who desire to advance technology education collectively. Such a global federation could do much toward providing educational leadership through a series of targeted activities. Our field must advance from merely the sharing of knowledge to the formation of projects and activities that will lead education internationally. Such coordination will not be easy. But if we are to be the strongest, most active, and most exciting profession in the years to come, such coordination will be essential. The major objective for the federation would be to foster greater cooperation among technology educators worldwide by:

(1) providing networks and information exchange systems needed to encourage professionals everywhere to be involved with colleagues in other countries,
(2) disseminating examples of successful programs, activities, and research to other nations through communication systems,
(3) providing leadership through selected professional development programs,
(4) building databases designed to support technology teachers and their accomplishments on an international level, and
(5) recognizing the achievements of dedicated educators and assisting in transferring those successes to other locations.

The idea of a global federation in education is not new. The science and reading skills communities have been participating in such groups for some time. Technology educators have flirted with this idea for years, but have never successfully created a structure that goes beyond the process of updating colleagues on the latest practices and changes evolving in each other's countries. Therefore, work needs to start immediately to create an organizational structure to satisfy desired needs, purposes, and goals that will move the profession in a positive direction. The responsibility for the creation of such a group should be with every national technology education organization with any interest in global interaction.

The creation of such a federation could most effectively be built using each country's national technology education organization as one of the components of the working structure and delivery system. This will allow for work to be completed and disseminated through the existing national structures, rather than building new communication networks in each nation. Therefore, the information flow and the effectiveness of a federation would be enhanced as a result of national structures that are already in place. There is no doubt that emerging technologies and the reality of change will continue to generate a demand for more effective communication in our profession. Our competition is not each other, but ourselves, as we have the challenge of producing the most effective educational system possible. This must be done in conjunction with the political and social institutions that exist in our own countries. We are on the threshold of providing leadership to the educational community by developing a worldwide federation, a federation which would be dedicated to advancing the development of technology education on an international level. No one can do it but us! No one can do it better than us!

Commonalities

Any effort to foster greater cooperation among technology education colleagues must contain a certain set of common beliefs and goals. Such understandings must relate to the philosophy and ideals of education, the field of technology education, and the teacher/student environment. The following commonalities are offered as an example or starting point for further discussion. Commonalities pertaining to the overall direction of technology education in a federation might include:

- allowing for far-reaching relationships with other subject areas
- serving as a means of developing knowledge
- examining technological alternatives to societal problems
- providing an opportunity to explore change
- exploring the future.

Further, a federation may help the field by enhancing and communicating successes with the following selected thinking skills as they relate to the teacher/student environment. These are:

- making decisions about technology
- learning how to learn
- developing manipulative skills and problem-solving techniques
- learning sources of information
- learning how to acquire information
- learning how to put information and materials into construction activity
- developing social skills for lifelong learning.

The strength of technology education lies within the expertise of the teachers in the classroom and laboratories. With this thought in mind, a federation should work to promote the following selected characteristics and inspire excellence in technology teaching. The teacher should:

- have faith in youth's intelligence
- see education as a process-not splintered exercises
- emphasize procedures, inquiry, and discovery
- see role as stimulator, guide, evaluator, and facilitator
- have a positive approach
- stimulate student refinement
- stay informed of technological developments
- have a broad perspective of possible student tasks.

With these thoughts in mind, a federation could go about the business of completing the five suggested goals outlined as a part of the vision in this presentation.

Action

The development of a worldwide technology education federation will result from initiative, coordination, and planning. If the ideas in this paper are considered to have merit for the profession, then the following procedures should be considered.

1. A directory of organizations with national status and possible interest in a federation should be developed for future communication purposes.
2. A letter of inquiry should be sent to potential association members of the federation.
3. Assuming a positive response from potential participants, a date should be established to hold an organizational meeting. That date should be announced at least 18 months in advance.
4. An agenda will need to be created to conduct the business of the first meeting, which would primarily address the mode of operation, organizational documents, and directions and funding for the group.

Subsequent meetings of the federation could be structured to increase the working effectiveness of the group. Examples of topics to be addressed would include establishing a clearinghouse, formalizing databases, creating professional development programs, and streamlining communication practices. Many barriers will be encountered during this process. Financial support will be the first major obstacle to such an undertaking. From the beginning, the success and degree of impact that a federation may have will depend upon funding sources. It will quickly become everyone's task to seek support from multinational corporations, agencies, or foundations. Initial meetings will possibly be held in settings with much of the expense being covered by the participant's national association or from personal contributions. However, with any initial vision or dream, determination on behalf of the creators will produce long-term success.

Summary

The preceding discussion aimed at fostering greater cooperation among technology education professionals has intentionally gone beyond the thought of having a series of meetings held at random locations over a period of time. We are in an age where we must live with and acknowledge great uncertainty. We must embrace error as a result of a fast-moving, highly sophisticated, technological society. If we are to be successful educators, we must evaluate the present in light of anticipated futures. We must be open to changes in commitments and directions.

Therefore, the time has come to create an international federation that serves as a clearinghouse for ideas and data, promotes commonalities that will advance the profession, and allows for dialogue to further thought and practice. The creation of such a federation will provide an avenue for developing one of the stronger curricular areas within our educational system and create support for innovative ideas to be transferred on an international basis.

The idea for such a federation is not new. Other educational fields are already attempting similar relationships with their colleagues. If technology educators are to keep pace with fellow educators, they will need to develop a consistent, concerted effort to advance the field at the international level. If technology educators are to keep pace with each other and foster greater cooperation in the development of the field, then an organized, structured federation is a feasible way to provide the coordination and direction to make our field one of exemplary quality at the international level.

Technology Education in the National Curriculum in England and Wales

Paul Black

King's College, UK

Abstract: The newly established National Curriculum in England and Wales includes technology as one of ten subjects that all pupils are required to study between the ages of 5 and 16 years. The curriculum is specified in detail, with emphasis on the process of tackling technological problems, comprising four stages of Identifying Needs and Opportunities, Designing, Making, and Evaluating. The scope is to be broad, including activities previously pursued separately as craft, or design, or home economics, or business studies, or applied science in engineering. The problems of implementing this new curriculum are discussed.

Key words: national curriculum, design, curriculum, learning progression, values, craft, science

A New Curriculum for Technology

The National Curriculum for England and Wales, established by the Education Reform Act of 1988, comprises ten subjects which, together with religious education, must be taught to all pupils [9]. Technology is one of these, and a detailed specification for this subject covering the age range 5 to 16 was published in 1989 [6]. From September 1990, pupils in the first year of the age range 5 to 7 (called Key Stage One) and pupils in the first year of the age range 11 to 14 (called Key Stage Three) are to start work in the subject as newly defined, and the whole program will be phased in for all pupils in the next few years [3].

Each of the new national curriculum subjects is specified according to the same general format. First, there is a division into a few broad areas, known as Profile Components; in the case of technology there are two such components, Design and Technology Capability and Information Technology Capability. In this paper, I shall discuss only the first of these.

Each component is then specified in terms of Attainment Targets (AT's) and Programme of Study. The former specify the objectives for pupils' learning, the latter set out "the essential matters, skills, and processes which need to be covered by pupils." Design and Technology Capability comprises four Attainment Targets as follows:

NATO ASI Series, Vol. F 78
Integrating Advanced Technology into Technology Education
Edited by M. Hacker et al.

AT One -- Identifying needs and opportunities
AT Two -- Generating a design
AT Three -- Planning and making
AT Four -- Evaluating.

Within each of these targets, the details are set out as a set of Statements of Attainment. Such statements are specified in ten levels which define the sequence of attainment which pupils should be helped to achieve. For example, level two statements are set as targets for the average 7-year-old, level three for the average 9-year-old, level six for the average 15-year-old, level nine for the exceptional 15-year-old. Each AT includes between 24 and 30 statements; examples will be given later in this paper.

The Programmes of Study are set out separately for each of the four Key Stages. All four include the following points:

- pupils should design and make artifacts, systems, and environments in response to needs and opportunities identified by them;
- they should work in the contexts of home, school, recreation, community, business, and industry;
- they should draw upon knowledge and skills in other subjects, particularly science, mathematics, and art;
- their work should also involve a growing understanding of needs and beliefs of people, in other cultures and in the past as well as in the present;
- they should progress through having increasing opportunities to identify their own tasks, and should make products which are more complex or which satisfy more demanding needs;
- pupils should discuss with each other and work both individually and in groups;
- pupils should work with a range of materials including textiles, graphic media, construction materials (such as clay, wood, plastic, metal), and food.

A third and final aspect of the curriculum will be the arrangements for national assessment. These have not yet been specified for technology. A general rationale for such assessment has been set out. It laid stress on the importance of formative assessment. The system of ten levels within each AT was meant to develop this aspect by expressing the curriculum in terms of pupil progression [4,5].

Tasks and Resources

The outstanding feature of this curriculum is that it embodies a new vision. There is a danger that teachers already dealing with some of its aspects in a partial way will say "we're doing this already," missing the point -- that the range and quality of aims and of experiences is an essential feature of what is indeed a radically new vision. A statement which makes clear the breadth of this vision was formulated by the U.K. Royal Society of Arts as part of an "Education for

Capability" movement [2,7]. This states: "There exists in its own right a culture which is concerned with doing, making and organizing in the creative arts. This culture emphasizes the day-to-day management of affairs, formulation of solution of problems, designing, manufacturing, marketing of goods and services." A little later it adds that "educators should spend more time preparing people in this way for life outside the educational system."

This sums up the main educational aim, which is to inject the new element of practical capability into the curriculum. Thus the argument here is that *design and technology* has a fundamental role to play in the curriculum. A related point is that in playing that role, it should be drawing in, and providing a new perspective for, other existing subjects in the curriculum.

That leads on immediately to questions about the second issue: who will actually teach this curriculum? Those who automatically yoke together "science and technology" will think that the main responsibilities will lie with science teachers. There is a long history of contention over this issue in Britain [10]. There are strong arguments, both in principle and in practice, why the two should not be so closely linked.

There are certainly physicists, for example, who regard technology as a small subset or offshoot of physics. They have to change this vision to one in which technology is a far more broad and extensive field than physics, with which it has a significant, but by no means major, overlap. All of the sciences, including physics, must be seen as contributors to a larger enterprise which has aims which extend well beyond those of science work and which therefore involves disciplines, resources, and skills from a very wide range.

Figure 1 is an attempt to show visually the relationship between knowledge in traditional subjects, *e.g.*, the conceptual knowledge of a subject such as science, with the very different and more complex area of real-world problems. It is possible to move between these two areas by two main steps. The one step is from theoretical science towards investigations in science, in which one uses specific knowledge and skills to pursue the activity of science. However, the scientist as scientist is interested in constructing and testing models of how the natural world behaves and can choose those problems which seem most likely to serve this purpose. In confronting real-world problems, the technologist is using knowledge and skills for a quite different purpose. That purpose is to meet a need defined by a customer or a situation. To respond, the technologist is usually forced to cross subject boundaries in order to bring in ideas and resources from other parts of the curriculum.

The idea can be expressed in a different way by saying that in technology, the *task* comes first. Usually, it cannot be amended to suit the *resources* most readily available. Instead, its needs dictate the range and scope of the resources to be employed. In developing a model of *tasks-with-resources* to characterize the role of technology in the school curriculum, Black and Harrison [1] argued that a school could only do justice to the nature of technology by using a cross-curricular approach to help pupils develop practical capability through tackling tasks.

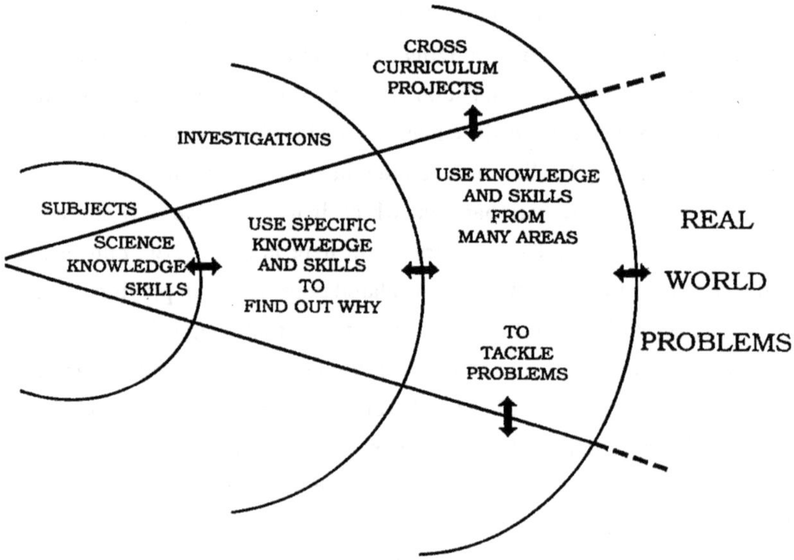

Figure 1. A model to relate real-world problems to subject knowledge

A Cross-Curricular Approach

Once the importance of a cross-curricular approach is accepted, there arises in schools in England and no doubt elsewhere, practical difficulties about who might teach the newly defined subject. It is being assumed in England that the scope of the subject as now defined means that teachers who previously taught craft design and technology, home economics, business studies, and art and design will have appropriate skills, and will have lost the bulk of the former work in their own specialties. There are several dangers here.

The first danger arises from the fact that established traditions may lead to design and technology being put in the hands of the craft design and technology (CDT) teachers. Such teachers may be charged with unifying command of the area and might claim that they know all about the subject. The home economics teachers, who also have a contribution to make, may then feel downtrodden and able only to cope with a fraction of the subject. The business studies teachers may feel the same, while teachers of art and design may withdraw from an uncomfortable liaison and concentrate on teaching art for its own sake.

A second and related danger is that these staff will divide up the work among themselves and then carry on independently in such a way that pupils will experience a wide range of different styles of learning and discern no unity in the subject. The teaching may merely cycle around the areas home economics, CDT, business studies, art and design, using a "carousel" approach in which the separate teachers teach in their four separate areas and the pupils go round them in sequence. That could be disastrous because well-designed tasks and a good balance across tasks cannot be achieved by work isolated within the boundaries that currently exist. *Design and technology* is more than the accumulation of these four separate subjects.

A third danger is that such a group will not feel the need to make an active link with other curriculum areas in their school, so that the importance of using (say) science teachers as resource persons for pupils pursuing technology tasks will never be recognized.

It will be difficult to realize the new vision because it involves radical changes in perceptions. A further problem is that it also involves, for individual teachers in schools, threats to their status, threats to their territory, threats to the resources they "own," threats to their promotion. These are serious issues, and finding effective ways to overcome them and achieve the vision is a challenge which will confront all schools for a long time to come.

One feature that is evident from studies that my colleagues and I have made in promoting cross-curricular technology in schools [11] is that good things do not happen in that difficult situation unless senior school management takes a very positive planning role in working out and directing a school policy to achieve full collaboration. If that role is not taken positively, and if the problems are left for subject teachers to resolve among themselves, then it seems unlikely that satisfactory solutions will be achieved.

The problems of implementing a cross-curricular approach may look quite different in a primary school. Primary teachers are not subject specialists and are used to working with tasks which are real to the children and which are treated in holistic fashion. The problem for them is to develop the skill of distilling from such activities the opportunities to develop particular pieces of science knowledge, mathematics knowledge, English knowledge, and so on. Thus, the design and technology activity required at primary level can provide a framework, a focus, for activities in which children can be engaged and through which their science, mathematics, and other knowledge can be explored, but where a broader look at problems as a whole is given priority.

A Curriculum of Tasks

The curriculum aims have to be turned into reality through the classroom experiences of pupils. In order to illustrate the issues involved here, it is useful to consider an imagined set of such experiences. The following gives a brief outline of the series of practical tasks which a pupil might experience in design and technology. The sequence envisaged is for the first year of work for a pupil, called Max, who has just entered a secondary school at age 11 [12].

Task 1. Looking at drinking mugs, how they are designed, how they are made, and evaluating them; serves AT's One and Four.

Task 2. After history lessons had been about the local community, surveying local buildings and trying to evaluate their design and to understand their fitness for their purpose and reasons for choosing the materials used to construct them; serves AT Four.

Task 3. Reviewing procedure for new entrants to the school which Max had recently experienced; the aim was to design a better system for new pupils for next year; serves AT Four leading to AT's One and Two.

Task 4. Designing and performing a puppet show where the pupils had to think about the music, which was done with computer composition, and about the script and presentation, which was done with computer word-processing; this led, for Max, to making one of the puppets; serves AT's Two, Three, and Four.

Task 5. Looking at the winter diet of old people, talking to some old people about needs and preferences and then designing, preparing, and evaluating a suitable meal; serves all four AT's.

Task 6. Evaluating winter garments, leading to some work on a new design but not to actual making of a garment; then looking at advertisements for such garments and composing alternatives using desktop publishing, to see how well they could look and how effective they might be; serves AT's Four, One, and Two, then all four AT's.

Task 7. Evaluating activity toys for use in the open, leading in to designing and making different toys; Max worked on a space capsule, going right through the designing, making and evaluating sequence; serves all four AT's.

Task 8. Studying a local park in order to design improvements to the play facilities for young children; taking part of the pond in the park and redesigning it as a sailing boat area complete with safety warning notices; serves all four AT's.

If such a scheme were to be proposed as a plan, requiring scrutiny or audit before adoption, or were to be carried out, so requiring evaluation during and after its implementation, the first need would be to define the criteria against which its effectiveness might be judged.

Curriculum Audit

One clear criterion would be the balance of the set of tasks as a whole, over the AT's and the Programmes of Study (POS). Some of the tasks mentioned above did not involve all four AT's. They involved, for example, some evaluation, then identifying needs and opportunities, and stopped. Are such incomplete tasks a useful activity? Some teachers would rule them out because they believe that full educational value depends on covering the whole cycle of activity, as represented by all four AT's

Other aspects of balance are brought out by reviewing the POS's. An obvious criterion is the requirement to explore artifacts and systems environments. Max's experience covered all three. Furthermore, at least some of the activities arose from needs and opportunities identified by the pupils themselves. There was also a range of contexts: the school, the community, and retail business were covered, but there was nothing involving industry; an audit process could lead to pressure to include an industry based activity.

Another related issue would be the range of knowledge and skills required by the tasks and the level to which these resources were being used. The Programmes of Study can only spell out in general terms the knowledge and skills required or assumed. However, positive efforts will have to be made to ensure that knowledge and skills from science and mathematics are substantially used in some of the projects. If they are not used then both a wrong impression of technology will be given and the opportunity to integrate the curriculum will have been missed.

Progression

The curriculum was designed to reflect and to support the ways in which children should progress. Learning progression can be seen both in different aspects of the POS's, and in different aspects of the AT's. To consider this more closely, here is set of statements taken from the AT concerned with Identifying Needs and Opportunities (AT One).

LEVEL 2 -- Ask questions which help them to identify the needs and opportunities for design and technological activity.

LEVEL 4 -- Starting with an unfamiliar situation, identify needs and opportunities for design and technological activity.

LEVEL 7 -- Analyze information of several kinds and draw conclusions about the needs and opportunities for a design and technological activity, recognizing and resolving conflicting considerations about what is worth doing.

While the first one, at level 2, is simply to "ask questions," the next one says "starting with something unfamiliar" -- that has made it more difficult. There is both a progression to the unfamiliar and a higher level expected -- more initiative from the pupils in identifying needs and opportunities. "Analyze information of several kinds and draw conclusions," "recognizing and resolving conflicting considerations" indicate that it is more demanding still: a variety of resources should be drawn upon, and the question of optimization in resolving conflicting constraints now comes in

Opinions about different varieties of progression are also to be found among the POS's; several aspects can be extracted. One aspect is that of the context, where this might progress from the personal and the local to the national and the global.

A second aspect of learning progression is the range. As children get older they should cover increasingly the whole range of industry, business, leisure, the home.

The third aspect is progression of knowledge and skills. Pupils should work from commonsense ideas, and contrivance technology methods, to using deeper concepts from a variety of subjects and higher levels of technique.

A fourth aspect of progression is in time-scale and difficulty. Young pupils can concentrate for a short time, older ones can be expected to exercise more persistence and to work over a longer time. They can also be expected to progress in meeting difficulties, in grappling with frustrations by themselves rather than being helped out of them, in showing inventiveness. So time-scale and difficulty are connected.

The fifth aspect is constraints. Tasks may start off being simple, rather artificial, and having few constraints. They would go on eventually to being complex, more real, and with multiple constraints.

A sixth aspect of progression may be in collaboration. There are very interesting statements in the POS's about leading a team and about organizing work among a team at Key Stage Four (ages 14 to 16), whereas what is implied at Key Stage One (ages 5 to 7) is just individual activity or voluntary collaboration. So from working as individuals to working in smaller groups to working in a syndicate or team approach is progression in pupils' ability to collaborate. Such progression will link to development of pupils' powers of communication. A syndicate working on different aspect of the one task soon finds that good communication is essential.

And finally, progression in learning is spelled out at different levels. I want to illustrate that point by selecting some further statements from AT One, as follows:

LEVEL 4 -- Know that in the past and in other cultures people have used design and technology to solve familiar problems in different ways.

LEVEL 5 -- Recognize that economic, social, environmental, and technological considerations and the preferences of users are important in developing opportunities.

LEVEL 6 -- Explain how different cultures have influenced design and technology, both in the needs met and the opportunities identified.

LEVEL 8 -- Investigate how needs and opportunities have led to design and technological activity in other cultures.

In this sequence, the target starts at the lowest level with "know that" then "recognize that" which is an advance on "know that." "Explain how" is more demanding again -- we are now digging deep for an explanation of how it was all put together, working still on technology in other times and other cultures. For the last statement, "investigate how" demands even more. Pupils have to dig into it and build up the story for themselves.

I have not exhausted the varieties of progression. It is important to look carefully in this way because one has to dig below the surface and develop some analytic skill for these statements in order to discern the ways in which they incorporate progression for guiding children's work.

In any audit, in deciding whether particular tasks are suitable, one key issue will be: are these tasks suitable in matching to and moving forward children's progression in these various aspects?

Other Times, Other Cultures

The extracts quoted above from AT One show how this aspect features in the curriculum. There are particular problems about meeting the requirements here. Appropriate resources, at least of information, for pupils to study how technology was implemented in other times and other cultures are not readily available and some way of providing those resources to schools has to be found. Work here will be in the form of case studies rather than practical tasks. Such case studies will require a type of classroom work quite different from that needed for open tasks.

Awareness of Values

The discussion of values can start by considering whether we are talking merely of knowing about uses and effects of technology. The curriculum specification contains many statements about effects: economic effects, social effects, judgments of value, choices, and so on. This means giving value to and arguing about the values implied by different choices. Pupils should be asking, "Who can make these choices" and "How do we make them?" How do we get to the point of recognizing that the technology does not happen inevitably, that it does happen by people choosing to do things?

However, there are then some difficult problems. In an evaluation of schemes within the UK's Technical and Vocational Initiative, David Layton [8] found several ways in which schools evade the issues, for example, teachers who were used to working in "hard edged" subjects like science and CDT would say that the discussions can only be vague and that they cannot see any definite and worthwhile outcome. Pupils say that there is nothing to "learn;" they can make things and they can learn things by heart, but discussions ending with unresolved differences do not seem to them to achieve anything. Behind such statements lie assumptions about maturing and about what constitutes learning which are not recognized by some teachers but are very familiar to other teachers. What may look to some teachers and pupils like an inconclusive discussion which did not get anywhere may in fact be an occasion in which pupils expanded their insight, moved away from prejudices, and went away seeing that there was far more to an argument than they, in their summary quick judgments, had imagined. Real learning can be going on and an inexperienced teacher not used to working in such areas can fail to recognize it.

There is also danger, of course, of political bias and controversy coming into the classroom. Those dangers have to be faced, although, clearly, there has to be very careful management by teachers in handling varieties of opinion. There is also the notion that it is all to do with "them" out there, that we are not responsible, so that the study of technological issues loses its point.

And finally there are difficulties associated with the notion of rapid change. Some of the school topics that have been used can come up in the daily news and make it hard to follow the teaching program that is envisaged.

The whole area is fraught with problems with few neat solutions. One way is to start with personal links: to work with issues that affect, say, particular groups like handicapped people and elderly people, which can be effective if pupils actually talk to such people and realize how solutions which appear appropriate to the pupils do not work, because when you start talking to people themselves, you have to think again about what works for them. Thus there is a very important first lesson to be gained by that direct contact. Perhaps we should also work at local issues which are more manageable and understandable and their effects more visible, particularly in studying, say, local industry and its effects on employment and its needs from the region.

Issues of historical perspective may also have particular value here. You can stand back and look at and study what has happened, and particularly the consequences of decisions. In schools where humanities staff were taking part in technology, it was found that this type of work could progress to a better level because such staff had better personal and social skills in handling such issues than science, technology, and maths staff.

One feature of this area ought to be clear. To avoid the issues concerning beliefs and values is to decide to teach a powerful lesson: that they are not important, they are not manageable, they are personal opinion only. This could be a very powerful, and powerfully bad, lesson about technology as a whole.

Pupils

The position of pupils is an issue which deserves more space than I can give here. I shall mention only three issues. One is that the presentation of tasks to pupils is crucial. I have certainly seen, in some open-ended tasks we have studied both in science and in technology, that the precise form of words, the precise way in which the target is expressed to pupils, can make a lot of difference in helping them to be very clear as to what they are supposed to do and how they might get on with it. When the teacher's specification is careless and not thought through, pupils get quite the wrong idea and chase the wrong target. Also, by presenting the same task in different ways, a teacher can expand the demands it makes or contract them. This "tuning" to adapt to the needs of pupils is a crucial skill.

Secondly, the way in which pupils are to be guided as they pursue a task can present very delicate dilemmas. Again, my colleagues and I have seen that some teachers, in supervising open-ended work, make the mistake of saying that because the task is "open" they will give no guidance. They then leave the pupils entirely on their own, not interrupting them, leaving them often to do some very strange things. This seems to me quite wrong and even rather silly. On the other hand, we have seen situations in which pupils have got so far and have then been stuck and the teacher has intervened to direct their attention to some resources, to some skills, to some ideas they had not thought of; then they take off and the thing flowers from that point. Skillful guidance, which gives neither too much direction nor too little, is essential.

And finally, reflection. Pupils are very ready, particularly when they are new to this type of work, to dash in, do something, and display the result, perhaps with pride. Very rarely however are they willing to spend time thinking about the task beforehand or willing to sit and reflect about what they have done afterwards. If, in addition, the lesson times are so designed that there is little time anyway for thoughtful reflection, some of the most important lessons will be lost.

There would be an important difference between Max's having had no more than an interesting set of experiences and Max's having developed capability in technology through those experiences. Reflection by pupils about what they have done and their involvement in the assessment of their work might be effective means to achieve this distinction, which is an essential target if we are to help pupils develop capability.

References

1. Black P.J. and Harrison G.B. *In Place of Confusion*. London, England: Nuffield Chelsea Curriculum Trust and the National Centre for School Technology, Trent Polytechnic, 1985.
2. Cross A. and McCormick R. (eds.). *Technology in Schools.* Milton Keynes, England: Open University Press, 1986.
3. Department of Education and Science. *National Curriculum - From Policy to Practice.* London: DES, 1989.
4. Department of Education and Science. *Task Group on Assessment and Testing -- A Report.* London: DES, 1988.
5. Department of Education and Science. *Task Group on Assessment and Testing -- Three Supplementary Reports.* London: DES, 1988.
6. Department of Education and Science. *Technology in the National Curriculum.* London: Her Majesty's Stationery Office, 1990.
7. *Education for Capability Movement Promoted by the Royal Society for the Encouragement of Arts Manufactures and Commerce*, 8 John Adam Street, London WCN 6E, England (various publications).
8. Layton D., Medway P., and Yeomans D. 14 - 18 *The Range of Practice - The T.V.E.I. Initiative.* Sheffield, England: The Training Agency, 1989.
9. MacClure, S. *Education Re-Formed: A Guide to the Education Reform Act 1988.* London: Hodder and Stoughton, 1988.
10. McCulloch G., Jenkins E. and Layton D. *Technological Revolution.* Lewes, England: Falmer Press, 1985.

11. Murray R. (ed.). *Managing Design and Technology in the National Curriculum.* Oxford, England: Heinemann Educational Books, 1990.

12. National Curriculum Council. *Non-Statutory Guidance for Design and Technology.* York, England: National Curriculum Council, 1990.

Summary Recommendations

The NATO community of countries shares a culture and a society that is increasingly dependent upon technological endeavor. The citizens of these nations, in order to be adequately prepared to live fulfilling and productive lives in the decades ahead, need to evaluate, control, and function confidently with the technology that characterizes their environment.

This conference, comprising governmental representatives, industrialists, educators, and academics, believes that, in an era characterized by a rapidly increasing rate of technological change and expansion of knowledge, all citizens need an education which develops their technological understanding and capability, and thereby enhances the cultural, social, and economic well-being of their societies. Therefore the conference recommends:

1. the establishment in all NATO countries, for all students throughout their compulsory schooling, of technology education programs, which have the following intents:
   a. to provide citizens with the capability to analyze and make value judgments about socio-technological issues and problems,
   b. to provide people with the capability to take informed action appropriate to meet technological challenges and exploit technological opportunities,
   c. to develop the understanding of and ability to cope with the technological influences upon their lives,
   d. to provide a broad basis of technological experiences so as to enable students to make informed career decisions,
   e. to provide a core of technologically creative and competent individuals who can contribute to a human-centered technological development.
2. the development of policies to ensure appropriate preservice and inservice education of teachers with requisite knowledge, skills, and attitudes to implement such technology education programs.
3. the development of policies to ensure adequate resourcing for the delivery and implementation of the programs.
4. to study further the appropriate role of advanced technology for creating new learning environments and for changing the content of technology education programs.
5. the organization of a NATO Advanced Study Institute to conceptualize and disseminate approaches and models for technology education.
6. the organization of a NATO Advanced Research Workshop to examine the research issues related to the teaching and learning of technology education.
7. the establishment of a world federation for technology education to foster exchange of information, research, and practice as it relates to contemporary technology education worldwide.

List of Participants

DIRECTOR
Michael Hacker
New York State Education Department
One Commerce Plaza, Room 1619
Albany, New York 12234, USA

BELGIUM
G. Staelens
Markeveld St, 31
8100 Torhout
Belgium

CANADA
Dr. George Hache
Memorial University of Newfoundland
Division of Special Programs
St. John's, Newfoundland
A1B 3X8, Canada

Ms. Dominique Bloy
Executive Director Distance
Education and Technology
Manitoba Education and Training
1970 Ness Avenue
Winnipeg, Manitoba
R3J 0Y9, Canada

Graham Carr
Ministry of Education
Curriculum Policy Development Branch
16th Floor, Mowat Block
Queen's Park, Toronto, Ontario
M7A 1L2, Canada

Mr. Rudy Pilz
Lord Dorchester Secondary School
61 Queen Street
Dorchester, Ontario
N0L1G0, Canada

DENMARK
Peter Steen Jensen
Danish Ministry of Education
Frederiksholms Kanal 26
1220 Copenhagen K, Denmark

Vagn Lauersen
Direktoratet for Gymnasieskolerne
Amargertorv 14
1160 Copenhagen K, Denmark

FRANCE
Dr. Marc Baudry
Universite du Maine
Faculte des Sciences
Departement de Mathematiques et
d'Informatique
Route de Laval
B.P. 535, 72017 LeMans Cedex, France

Jean-Jacques Duby
Group Director of Science and Technology
IBM Europe S.A.
Tour Pascal
22, Route de la Demi-Lune, Puteaux
Cedex 40
92075 Paris-La Defense, France

Dr. Martial Vivet
Universite du Maine
Laboratoire d'Informatique
Route de Laval
B.P. 535, 72017 LeMans Cedex, France

GERMANY
Prof. Dr. Dietrich Blandow
Pedagogical University
Nordhauser Strasse 63
0-5064 Erfurt, Germany

GERMANY, cont'd
Dr. Gunter Bohle
Siemens AG, Semiconductor Group, HLOA
Balanstr. 73
W-8000 Munchen 80, Germany

Dr. Andreas Bottcher
BASF Aktiengesellschaft
Abteilung Hochschulbeziehungen
Postfach
W-6700 Ludwigshafen/Rhein, Germany

Dr. Wolf E. Traebert
Westfalische Wilhelms Universitat Munster
Institut fur Technik und Ihre Didaktik
Scharnhorststr. 100
W-4400 Munster, Germany

Dr. Gregor Tyrchan
Bergische Universitat Wuppertal
Fach: Technologie
Gaussstr.
W-5600 Wuppertal, Germany

GREECE
Dr. Stavros Androulidakis
Technological Education Institute
Thessaloniki
P.O. Box 14561
Thessaloniki 541 01, Greece

Dr. Constantinos Tsiantis
Technological Education
Institute of Athens
59 Granikou Street
16233 Vironas
Athens, Greece

GREECE, cont'd
Mr. George Kalkanis
Ministry of National Education
and Religious Affairs
15 Mitropoleos Street
Athens 10185, Greece

NETHERLANDS
Dirk Blom
Adj. Dir. P&A
Philips Nederland BV
Gebouw ECY 4
P.O. Box 218
5600 MD Eindhoven
The Netherlands

Dr. Feye Meijer
Philips Nederland BV
Gebouw WB5
P.O. Box 218
5600 MD Eindhoven
The Netherlands

Dr. Marc de Vries
Pedagogical Technological College
P.O. Box 826
5600 Av. Eindhoven
The Netherlands

Jenne van der Velde
National Institute for Curriculum
Development (SLO)
P.O. Box 2041
7500 CA Enschede
The Netherlands

Frans Timmermans
Pedagogical Technological College
P.O. Box 826
5600 Av. Eindhoven
The Netherlands

PORTUGAL
Maria Joao Alves
COURACA DE LISBOA
No. 34 30 ESQ
3000 Coimbra, Portugal

TURKEY
Dr. Fuat Bayrakceken
Ankara University
Department of Engineering Physics
06100 Tandogan
Ankara, Turkey

Dr. Yasar T. Hondur
Scientific and Technical Research Council of Turkey
Emek Ishani Kat 16
Kizilay 06420 Ankara, Turkey

Fatih Mehmet Sahin
TUBITAK
The Scientific and Technical Research Council of Turkey
Emek Ishani Kat 16
Kizilay 06420 Ankara, Turkey

Professor Emrah Orhun
E.U. Bilgisayar Arastirma ve Uygulama Merkezi
Bornova, Izmir, Turkey

Dr. Ugur Dilmen
TUBITAK
Istanbul CAD #88
Iskitler-Ankara
Turkey

UNITED KINGDOM
Anthony Gordon
Education Inspectorate
The Kingston Centre
Harrowby Street
Stafford, ST16 3TU
United Kingdom

Giles Job
Science Inspector
Education Inspectorate
The Kingston Centre
Harrowby Street
Stafford ST16 3TU
United Kingdom

Robert McCormick
School of Education
The Open University
Walton Hall
Milton Keynes
MK7 6AA
United Kingdom

Professor Paul Black
Education Department
Kings College
552 Kings Road
London SW3
United Kingdom

Edgar Jenkins
School of Education
University of Leeds
Leeds LS2 9JT
United Kingdom

UNITED STATES
John Gunn
AT&T Bell Laboratories
200 Laurel Avenue
Room 3P-170A
Middletown, New Jersey 07748

Dr. Daniel Householder
Department of Industrial, Vocational, and Technical Education
Texas A&M University
Harrington Education Center, Room 620
College Station, Texas 77843-3256

Dr. John Hutchinson
Trenton State College
Armstrong Hall
Hillwood Lakes CN4700
Trenton, New Jersey 08650-4700

Dr. John Larkin
School of Technology
Department of Technology and Vocational-Technical Education
Central Connecticut State University
New Britain, Connecticut 06050

Dr. Thomas Liao
Department of Technology and Society
College of Engineering and Applied Sciences
SUNY at Stonybrook
Stonybrook, New York 11794

Dr. Ernest Savage
College of Technology
Bowling Green University
Bowling Green, Ohio 43403

UNITED STATES, cont'd
Dr. Kendall Starkweather
Executive Director
International Technology Education Association
1914 Association Drive
Reston, Virginia 22091

Dr. Dennis Swyt
National Institute of Standards and Technology
Bureau Drive
Building 220, Room A109
Gaithersburg, Maryland 20899

Dr. Ronald Todd
Executive Editor Ties Magazine
College of Design Arts
Drexel University
Philadelphia, Pennsylvania 19104

NON-NATO PARTICIPANTS

AUSTRIA
Johann Ussar
VS II
Schulweg 1
A-8130 Frohnleiten, Austria

ISRAEL
Dr. Uri Zoller
Haifa University, Oranim
Division of Chemical Studies
Department of Science Education
P.O. Kiryat, Tivon
Israel 36910

Subject Index

NATO ASI Series F

Including Special Programmes on Sensory Systems for Robotic Control (ROB) and on Advanced Educational Technology (AET)

Vol. 1: Issues in Acoustic Signal – Image Processing and Recognition. Edited by C. H. Chen. VIII, 333 pages. 1983.

Vol. 2: Image Sequence Processing and Dynamic Scene Analysis. Edited by T. S. Huang. IX, 749 pages. 1983.

Vol. 3: Electronic Systems Effectiveness and Life Cycle Costing. Edited by J. K. Skwirzynski. XVII, 732 pages. 1983.

Vol. 4: Pictorial Data Analysis. Edited by R. M. Haralick. VIII, 468 pages. 1983.

Vol. 5: International Calibration Study of Traffic Conflict Techniques. Edited by E. Asmussen. VII, 229 pages. 1984.

Vol. 6: Information Technology and the Computer Network. Edited by K. G. Beauchamp. VIII, 271 pages. 1984.

Vol. 7: High-Speed Computation. Edited by J. S. Kowalik. IX, 441 pages. 1984.

Vol. 8: Program Transformation and Programming Environments. Report on a Workshop directed by F. L. Bauer and H. Remus. Edited by P. Pepper. XIV, 378 pages. 1984.

Vol. 9: Computer Aided Analysis and Optimization of Mechanical System Dynamics. Edited by E. J. Haug. XXII, 700 pages. 1984.

Vol. 10: Simulation and Model-Based Methodologies: An Integrative View. Edited by T. I. Ören, B. P. Zeigler, M. S. Elzas. XIII, 651 pages. 1984.

Vol. 11: Robotics and Artificial Intelligence. Edited by M. Brady, L. A. Gerhardt, H. F. Davidson. XVII, 693 pages. 1984.

Vol. 12: Combinatorial Algorithms on Words. Edited by A. Apostolico, Z. Galil. VIII, 361 pages. 1985.

Vol. 13: Logics and Models of Concurrent Systems. Edited by K. R. Apt. VIII, 498 pages. 1985.

Vol. 14: Control Flow and Data Flow: Concepts of Distributed Programming. Edited by M. Broy. VIII, 525 pages. 1985.

Vol. 15: Computational Mathematical Programming. Edited by K. Schittkowski. VIII, 451 pages. 1985.

Vol. 16: New Systems and Architectures for Automatic Speech Recognition and Synthesis. Edited by R. De Mori, C.Y. Suen. XIII, 630 pages. 1985.

Vol. 17: Fundamental Algorithms for Computer Graphics. Edited by R.A. Earnshaw. XVI, 1042 pages. 1985.

Vol. 18: Computer Architectures for Spatially Distributed Data. Edited by H. Freeman and G.G. Pieroni. VIII, 391 pages. 1985.

Vol. 19: Pictorial Information Systems in Medicine. Edited by K. H. Höhne. XII, 525 pages. 1986.

Vol. 20: Disordered Systems and Biological Organization. Edited by E. Bienenstock, F. Fogelman Soulié, G. Weisbuch. XXI, 405 pages. 1986.

Vol. 21: Intelligent Decision Support in Process Environments. Edited by E. Hollnagel, G. Mancini, D.D. Woods. XV, 524 pages. 1986.

NATO ASI Series F

Vol. 22: Software System Design Methods. The Challenge of Advanced Computing Technology. Edited by J.K. Skwirzynski. XIII, 747 pages. 1986.

Vol. 23: Designing Computer-Based Learning Materials. Edited by H. Weinstock and A. Bork. IX, 285 pages. 1986.

Vol. 24: Database Machines. Modern Trends and Applications. Edited by A.K. Sood and A.H. Qureshi. VIII, 570 pages. 1986.

Vol. 25: Pyramidal Systems for Computer Vision. Edited by V. Cantoni and S. Levialdi. VIII, 392 pages. 1986. *(ROB)*

Vol. 26: Modelling and Analysis in Arms Control. Edited by R. Avenhaus, R.K. Huber and J.D. Kettelle. VIII, 488 pages. 1986.

Vol. 27: Computer Aided Optimal Design: Structural and Mechanical Systems. Edited by C.A. Mota Soares. XIII, 1029 pages. 1987.

Vol. 28: Distributed Operating Systems. Theory und Practice. Edited by Y. Paker, J.-P. Banatre and M. Bozyiğit. X, 379 pages. 1987.

Vol. 29: Languages for Sensor-Based Control in Robotics. Edited by U. Rembold and K. Hörmann. IX, 625 pages. 1987. *(ROB)*

Vol. 30: Pattern Recognition Theory and Applications. Edited by P.A. Devijver and J. Kittler. XI, 543 pages. 1987.

Vol. 31: Decision Support Systems: Theory and Application. Edited by C.W. Holsapple and A.B. Whinston. X, 500 pages. 1987.

Vol. 32: Information Systems: Failure Analysis. Edited by J.A. Wise and A. Debons. XV, 338 pages. 1987.

Vol. 33: Machine Intelligence and Knowledge Engineering for Robotic Applications. Edited by A.K.C. Wong and A. Pugh. XIV, 486 pages. 1987. *(ROB)*

Vol. 34: Modelling, Robustness and Sensitivity Reduction in Control Systems. Edited by R.F. Curtain. IX, 492 pages. 1987.

Vol. 35: Expert Judgment and Expert Systems. Edited by J.L. Mumpower, L.D. Phillips, O. Renn and V.R.R. Uppuluri. VIII, 361 pages. 1987.

Vol. 36: Logic of Programming and Calculi of Discrete Design. Edited by M. Broy. VII, 415 pages. 1987.

Vol. 37: Dynamics of Infinite Dimensional Systems. Edited by S.-N. Chow and J.K. Hale. IX, 514 pages. 1987.

Vol. 38: Flow Control of Congested Networks. Edited by A.R. Odoni, L. Bianco and G. Szegö. XII, 355 pages. 1987.

Vol. 39: Mathematics and Computer Science in Medical Imaging. Edited by M.A. Viergever and A. Todd-Pokropek. VIII, 546 pages. 1988.

Vol. 40: Theoretical Foundations of Computer Graphics and CAD. Edited by R.A. Earnshaw. XX, 1246 pages. 1988.

Vol. 41: Neural Computers. Edited by R. Eckmiller and Ch. v. d. Malsburg. XIII, 566 pages. 1988.

NATO ASI Series F

Vol. 42: Real-Time Object Measurement and Classification. Edited by A. K. Jain. VIII, 407 pages. 1988. *(ROB)*

Vol. 43: Sensors and Sensory Systems for Advanced Robots. Edited by P. Dario. XI, 597 pages. 1988. *(ROB)*

Vol. 44: Signal Processing and Pattern Recognition in Nondestructive Evaluation of Materials. Edited by C. H. Chen. VIII, 344 pages. 1988. *(ROB)*

Vol. 45: Syntactic and Structural Pattern Recognition. Edited by G. Ferraté, T. Pavlidis, A. Sanfeliu and H. Bunke. XVI, 467 pages. 1988. *(ROB)*

Vol. 46: Recent Advances in Speech Understanding and Dialog Systems. Edited by H. Niemann, M. Lang and G. Sagerer. X, 521 pages. 1988.

Vol. 47: Advanced Compuling Concepts and Techniques in Control Engineering. Edited by M. J. Denham and A. J. Laub. XI, 518 pages. 1988.

Vol. 48: Mathematical Models for Decision Support. Edited by G. Mitra. IX, 762 pages. 1988.

Vol. 49: Computer Integrated Manufacturing. Edited by I. B. Turksen. VIII, 568 pages. 1988.

Vol. 50: CAD Based Programming for Sensory Robots. Edited by B. Ravani. IX, 565 pages. 1988. *(ROB)*

Vol. 51: Algorithms and Model Formulations in Mathematical Programming. Edited by S. W. Wallace. IX, 190 pages. 1989.

Vol. 52: Sensor Devices and Systems for Robotics. Edited by A. Casals. IX, 362 pages. 1989. *(ROB)*

Vol. 53: Advanced Information Technologies for Industrial Material Flow Systems. Edited by S. Y. Nof and C. L. Moodie. IX, 710 pages. 1989.

Vol. 54: A Reappraisal of the Efficiency of Financial Markets. Edited by R. M. C. Guimarães, B. G. Kingsman and S. J. Taylor. X, 804 pages. 1989.

Vol. 55: Constructive Methods in Computing Science. Edited by M. Broy. VII, 478 pages. 1989.

Vol. 56: Multiple Criteria Decision Making and Risk Analysis Using Microcomputers. Edited by B. Karpak and S. Zionts. VII, 399 pages. 1989.

Vol. 57: Kinematics and Dynamic Issues in Sensor Based Control. Edited by G. E. Taylor. XI, 456 pages. 1990. *(ROB)*

Vol. 58: Highly Redundant Sensing in Robotic Systems. Edited by J. T. Tou and J. G. Balchen. X, 322 pages. 1990. *(ROB)*

Vol. 59: Superconducting Electronics. Edited by H. Weinstock and M. Nisenoff. X, 441 pages. 1989.

Vol. 60: 3D Imaging in Medicine. Algorithms, Systems, Applications. Edited by K. H. Höhne, H. Fuchs and S. M. Pizer. IX, 460 pages. 1990.

Vol. 61: Knowledge, Data and Computer-Assisted Decisions. Edited by M. Schader and W. Gaul. VIII, 421 pages. 1990.

Vol. 62: Supercomputing. Edited by J. S. Kowalik. X, 425 pages. 1990.

Vol. 63: Traditional and Non-Traditional Robotic Sensors. Edited by T. C. Henderson. VIII, 468 pages. 1990. *(ROB)*

Vol. 64: Sensory Robotics for the Handling of Limp Materials. Edited by P. M. Taylor. IX, 343 pages. 1990. *(ROB)*

Vol. 65: Mapping and Spatial Modelling for Navigation. Edited by L. F. Pau. VIII, 357 pages. 1990. *(ROB)*

NATO ASI Series F

Vol. 66: Sensor-Based Robots: Algorithms and Architectures. Edited by C.S.G. Lee. X, 285 pages. 1990. *(ROB)*

Vol. 67: Designing Hypermedia for Learning. Edited by D.H. Jonassen and H. Mandl. XXV, 457 pages. 1990. *(AET)*

Vol. 68: Neurocomputing. Algorithms, Architectures and Applications. Edited by F. Fogelman Soulié and J. Hérault. XI, 455 pages. 1990.

Vol. 69: Real-Time Integration Methods for Mechanical System Simulation. Edited by E.J. Haug and R.C. Deyo. VIII, 352 pages. 1990.

Vol. 70: Numerical Linear Algebra. Digital Signal Processing and Parallel Algorithms. Edited by G. Golub and P. Van Dooren. XIII, 729 pages. 1991.

Vol. 71: Expert Systems and Robotics. Edited by T. Jordanides and B. Torby. XII, 744 pages. 1991.

Vol. 72: High-Capacity Local and Metropolitan Area Networks. Architecture and Performance Issues. Edited by G. Pujolle. X, 536 pages. 1991.

Vol. 73: Automation and Systems Issues in Air Traffic Control. Edited by J.A. Wise, V.D. Hopkin and M.L. Smith. XIX, 589 pages. 1991.

Vol. 74: Picture Archiving and Communication Systems (PACS) in Medicine. Edited by H.K. Huang, O. Ratib, A.R. Bakker and G. Witte. XII, 445 pages. 1991.

Vol. 75: Speech Recognition and Understanding. Recent Advances, Trends and Applications. Edited by P. Laface and R. De Mori. 1991.

Vol. 76: Multisensor Fusion for Computer Vision. Edited by J.K. Aggarwal. 1991. *(ROB)*

Vol. 77: Computer Algorithms for Solving Linear Algebraic Equations: The State of the Art. Edited by E. Spedicato. VIII, 352 pages. 1991.

Vol. 78: Integrating Advanced Technology into Technology Education. Edited by M. Hacker, A. Gordon, M. de Vries. VIII, 185 pages. 1991. *(AET)*

MIX
Papier aus verantwortungsvollen Quellen
Paper from responsible sources
FSC® C105338

If you have any concerns about our products,
you can contact us on
ProductSafety@springernature.com

In case Publisher is established outside the EU,
the EU authorized representative is:
Springer Nature Customer Service Center GmbH
Europaplatz 3, 69115 Heidelberg, Germany

Printed by Libri Plureos GmbH
in Hamburg, Germany